T0301496

Climate Security and Climate Justice

Climate Security and Climate Justice
Recognizing Context in the Sahel

Tor A. Benjaminsen

Professor of Environment and Development Studies, Department of International Environment and Development Studies (Noragric), Faculty of Landscape and Society, Norwegian University of Life Sciences, Norway

Edward Elgar
PUBLISHING

Cheltenham, UK · Northampton, MA, USA

Published by
Edward Elgar Publishing Limited
The Lypiatts
15 Lansdown Road
Cheltenham
Glos GL50 2JA
UK

Edward Elgar Publishing, Inc.
William Pratt House
9 Dewey Court
Northampton
Massachusetts 01060
USA

A catalogue record for this book
is available from the British Library

Library of Congress Control Number: 2024941804

This book is available electronically in the **Elgar**online
Geography, Planning and Tourism subject collection
http://dx.doi.org/10.4337/9781035325184

ISBN   978 1 0353 2517 7 (cased)
ISBN   978 1 0353 2518 4 (eBook)

Printed and bound in Great Britain by
TJ Books Limited, Padstow, Cornwall

Contents

Preface

My first visit to Mali was in the dry and hot season of 1987. This was also my first fieldwork and visit to Africa south of the Sahara. The purpose was to collect data for a Cand. Scient. research project in Landscape Ecology and Resource Geography at the University of Oslo.

This fieldwork took place just after the big Sahelian drought of 1984–85 with all the global media attention that followed. It was also the time when a new concept – 'sustainable development' – was born, which represented a renewed interest in the environment in the midst of a neoliberal era. The answers to global environmental problems were seen to be more economic growth, not only in poor, but also in rich countries – a dominant paradigm that has continued until today, but more recently under the labels 'green growth' or 'ecomodernism' in the context of transformation to low-carbon societies.

This book builds on 37 years of recurrent research visits to Mali and it draws the lines and connections between the various themes that I have worked on over the years – desertification, pastoralism, land-use conflicts, Tuareg and jihadist rebellions. One main aim has been to discuss the background to the current Sahelian crisis. I argue that understanding this crisis requires a political ecology approach that combines historical and materialist perspectives with a focus on the context of people's motivations and agency. I also argue that 'sustainable development' forms the backdrop of the current jihadist rebellions in the Sahel.

While looking back at the causes of the current crisis, this book simultaneously looks ahead and engages with potential risks associated with ongoing debates about climate security and climate justice. Both these approaches tend to neglect the political ecological context in the Sahel. When misrecognition and flawed analyses guide policy interventions, there are clearly risks of adverse effects. Again, desertification re-emerges as a key policy theme, and the fight against this imagined problem becomes an essential part of climate action in the Sahel – in particular in the form of afforestation projects.

Then we come full circle, because climate mitigation through afforestation may not only lead to failed climate projects, but also to increased resistance especially from pastoralists who are dispossessed of their pastures, again potentially exacerbating conflicts.

In the 1980s and 90s, the main policy and media attention was on how to stop desertification. Farmers and pastoralists were thought to be both victims and villains. They were believed to exceed their own resource base through overpopulation and lack of knowledge about fragile ecosystems. The heroes were external (i.e. Western) experts who advised about issues such as sustainable range management (often leading to more conflicts and confusion) and forest reforms (promoting the privatisation of rights to harvest and sell wood often to the benefit of local elites with contacts in the Forest Service). To please international aid donors, Sahelian governments made legislation more 'environmentally friendly', which led to high fines for setting bush fires, for not having wood-saving stoves, or for collecting dry wood. The implication was an extra tax on the rural peasantry that served an urban elite.

In addition, in order to modernise agriculture, large- and medium-scale agricultural investments continued to dispossess farmers and pastoralists. Such investments were either initiated by the World Bank or other international donors, or emerged through internal political processes where politicians and high-level civil servants were able to grab land held under customary insecure tenure.

In other words, while the sustainable development hype led to more restrictions and taxes on rural people including resource and land dispossession, it also benefited political and bureaucratic elites through increased possibilities of rent capture. Over time, this resulted in widespread frustration and anger in rural areas throughout the Sahel – or 'grievances' in the vocabulary of peace and conflict studies.

While 'sustainable development' forms a backdrop to current jihadist rebellions, other factors that have played in are the failed democratisation processes in the Sahelian countries since President Mitterrand's famous speech to Francophone African presidents at La Baule in 1990[1] as well as the growth of a global jihadist movement since 9/11 and the subsequent invasions in Iraq and Afghanistan, and the NATO bombing of Libya.

Today, desertification is thought to primarily result from climate change leading to drier conditions, and afforestation has come back as a solution to stop this process. In addition, it has become one of the most relied upon climate measures to mitigate global climate change more generally. IPCC scenarios include enormous areas that need to be planted with trees to stay within 1.5- or 2-degrees global temperature increase. Most of this land is in Africa and, if implemented, such large-scale afforestation projects risk adversely affecting local livelihoods and increase land-use conflicts. Among Western governments, however, large-scale afforestation in Africa has become an enamoured climate mitigation measure, which takes pressure away from carrying out unpopular climate action at home. The idea is that through planting trees in Africa, the European or American way of life can be sustained.

In addition, it is often argued that climate change plays a key role as a driver or threat multiplier of conflicts in the Sahel – a view that I question in this book. This is not climate denialism. Climate change is real and extremely serious, and it affects the Sahel through rising temperatures and probably more irregular rainfall. This becomes a significant problem for farmers in particular, because it may be increasingly difficult to know when to put the seeds into the ground. But climate change also leads to more total annual rainfall in the Sahel and most probably not to a drying in contrast to the many claims about desertification and increasing resource scarcity.

I argue that emphasising climate change as a cause of the crisis in the Sahel is ahistorical and apolitical as it neglects the often complex historical and political contexts of various conflicts. Such a view simply takes attention away from the important political-economic-historical factors. In addition, it also ignores the research frontier in ecology and climate science. Much of the media attention, policy documents, as well as some of the research contributions to this mainstream view within climate security tilt towards Eurocentrism and colonial tropes about Africans without agency. Instead, Africans are seen to simply react to outside stimuli such as increased temperature or a change in rainfall patterns. The great Marxist historian E.P. Thompson found such context-less analyses reductionistic and paternalist (see chapter 3).

Taking context seriously does not only imply recognising Sahelian history and the agency of people, but it also means taking results from research in ecology and climate science seriously. This is not always the case within contributions to climate security and climate justice in the Sahel.

When I first arrived in Mali in 1987, I too believed that desertification caused by local over-exploitation was a serious problem in the area, and the purpose of my study was to investigate its causes. However, during fieldwork in the Gourma region in northern Mali, I observed large areas with dead trees after the drought three years earlier. There had, in other words, been massive deforestation and (temporary) desertification, but the dead forests were far from any village, meaning that it was not human over-cutting of wood that was the cause. The trees had simply died because of the drought.

While rainfall remained below average for a decade after the drought, recovery of the populations of woody plants had already started in 1985, which confirms the resilience of the vegetation in the Sahel.[2] Today, the density of trees in the Gourma is much higher than it was in 1987, and the grass cover is more abundant, although fluctuating with the annual rainfall.

During this first research stay in Mali, I was fortunate to be able to spend a few weeks in the field with Pierre Hiernaux, who worked as an ecologist at the International Livestock Centre for Africa (ILCA) at the time. Through this first field visit, as well as later ones, he patiently introduced me to some of the main characteristics of Sahelian ecology – such as instability and resilience.

Pierre also possesses an unusual time series of data on the vegetation in the Gourma, including photographs taken in the rainy season since the early 1980s and covering several decades, which illustrates both instability and fluctuations as well as long-term re-greening. His sophisticated ecological analyses, combined with detailed empirical data often including surprising results, became an eye-opener and changed my understanding of Sahelian ecology.

I am also grateful to André Marty (sociologist at IRAM) who allowed me to visit him at his home in southern France twice and taught me about pastoral production systems in the Gourma. During repeated field visits to northern Mali in the late 1980s and early 90s, I also learnt a lot from discussions with Sidi Mohammed Ould Zahaby (who later disappeared and turned up in 1992 as the spokesman for the rebel movement), Mohammed Ag Mahmoud (an autodidact researcher, school headmaster and local Tuareg leader) and Ibrahim Ag Youssouf (with a postgraduate degree in socio-linguistics from the US and respected resource person).

From 1989 to 1992, I worked as a research assistant at the Mali research programme at the University of Oslo. From this period, I am particularly grateful to Alida Boye who coordinated the programme, as well as the anthropologist couple Gunnvor Berge and Jon Pedersen. We had lots of fun together both academically and socially, in Norway and in Mali.

In Mali, I received practical and logistical support from Ousmane Sidibé (who later became minister in the government), Abdoulaye Touré, Drissa Diallo and Mahmadou Diallo Iam (director of the CNRST).

From 1994, I again had a chance to focus on Mali through a scholarship to do postgraduate research in Geography and International Development Studies at Roskilde University in Denmark, while being based at the Centre for Development and the Environment (SUM) at the University of Oslo. The intention had been to continue the work in northern Mali, but due to the Tuareg rebellion that had increasingly made larger areas unsafe after its start in June 1990, I decided to relocate field work and data collection to the cotton zone in southern Mali. While I revisit much of my work in northern and central Mali in this book, the research I carried out in the cotton zone in the 1990s is not included as it does not connect easily to the book's theme.

In 1998, I was lucky to receive an offer for a position as Associate Professor at the Department of International Environment and Development Studies (Noragric) at the Norwegian University of Life Sciences (NMBU) (then the Agricultural University of Norway). At Noragric I have, over the years, enjoyed the collegiality and friendship of many good colleagues that are too numerous to mention.

In 2022, I received an Advanced Grant from the European Research Council (ERC) for five years to continue my work in the Sahel, and in particular in Mali. The grant gave me the opportunity to employ two postdoctoral researchers (Ibrahima Poudiougou and Elieth Eyebiyi) and three postgraduate students (Tchello Kasse, Nengak Daniel Gondyi and Hannes Bräuer) within the Landresponse project (see landresponse.space). In addition, the project has two co-investigators in Mali – Boubacar Ba (director of the Centre for Analysis on Governance and Security in the Sahel) and Bréma Ely Dicko (professor and anthropologist at the University of Bamako). The aim of the project is to investigate the role of land dispossession in processes of migration as well as in the recruitment to armed groups in the Sahel.

I was particularly delighted that this new funding gave me the opportunity to continue a long-term cooperation with Boubacar Ba. We have known each other since 1997, and started to work together studying land-use conflicts in the inner delta of the Niger river from 2006. This partnership has developed into a fruitful research collaboration as well as a close friendship. Boubacar has an incredible network of informants and contacts in central Mali that I have benefitted tremendously from. What I think I know about people's motivations in central Mali, and in particular among Fulani pastoralists, I owe to Boubacar and the many people he has put me in contact with as interviewees over the years.

The approach of this book is also inspired by cooperation and conversations with a number of colleagues who have influenced my thinking on issues such as political ecology, climate security, climate justice, green transformation, peasant studies and the politics of the Sahel more broadly. They are in particular (in alphabetical order) Mariel Aguilar Støen, Tom Bassett, Piers Blaikie, Hannes Bräuer, Dan Brockington, Ian Bryceson, Halvard Buhaug, Bram Büscher, Connor Cavanagh, Frances Cleaver, Esteve Corbera, Elieth Eyebiyi, Bill Derman, Rosaleen Duffy, Anwesha Dutta, Jens Friis Lund, Denis Gautier, Nengak Daniel Gondyi, Erik Gómez-Baggethun, Sten Hagberg, Ketil Fred Hansen, Pierre Hiernaux, Randi Kaarhus, Tchello Kasse, Christian Lund, Faustin Maganga, Bill Moseley, Nancy Peluso, Ibrahima Poudiougou, Nitin Rai, Jesse Ribot, Paul Robbins, Melanie Sommerville, Espen Sjaastad, Sian Sullivan, Hanne Svarstad, Michael Watts and Poul Wisborg.

On a personal note, this book manuscript was written after I received a diagnosis of motor neurone disease in September 2023. After the first couple of months in shock, I discovered writing as a better remedy than anything to take attention away from a disease that gradually impairs my physical abilities.

I am immensely grateful to my family for all the support during these difficult times – particularly my dear wife Hanne and my wonderful daughters Eline and Mina, but also the broader family and friends, as well as colleagues. Eline has also helped me put together the list of references as well as drawing three figures and searching AI (see Box 1.1).

Finally, I appreciate permission received from journals and co-authors to reproduce shorter or longer excerpts from the following articles:

Benjaminsen, T.A. and P. Hiernaux, 2019. From Desiccation to Global Climate Change: A History of the Desertification Narrative in the West African Sahel, 1900–2018. *Global Environment* 12 (1): 206–236.

Benjaminsen, T.A. 2021. Depicting Decline: Images and Myths in Environmental Discourse Analysis. *Landscape Research* 46 (2): 211–225.

Benjaminsen, T.A. and B. Ba. 2021. Fulani-Dogon Killings in Mali: Farmer-Herder Conflicts as Insurgency and Counterinsurgency. *African Security* 14 (1): 4–26.

Benjaminsen, T.A., H. Svarstad and I. Shaw of Tordarroch. 2022. Recognising Recognition in Climate Justice. *IDS Bulletin* 53 (4).

Benjaminsen, T.A. and B. Ba. 2024. A Moral Economy of Pastoralists? Understanding the 'Jihadist' Insurgency in Mali. *Political Geography*.

Let me end with the words of the famous (and late) Malian desert blues musician Ali Farka Touré who visited my university NMBU in 1999. I had the honour to meet with him in my office, to have lunch with him and to show him around on campus. During our discussions about Mali, he repeatedly said: 'Ce que je dis n'engage que moi' – in other words I bear the sole responsibility for the arguments in this book and how they are articulated.

Oslo, 14th May 2024

Tor A. Benjaminsen

NOTES

1. The background to this speech was the end of the cold war with its political strategic implications in Africa. Mitterrand invited the African leaders to embark on a democratisation process or risk losing French support. This ushered in multi-party democracy in most Francophone West African countries within a couple of years.
2. Hiernaux, P., Diarra, L., Trichon, V., Mougin, E., Soumaguel, N., Baup, F., Woody plant population dynamics in response to climate changes from 1984 to 2006 in Sahel (Gourma, Mali), *Journal of Hydrology* (2009).

Acknowledgements

Funded by the European Union (ERC, LANDRESPONSE, project number 101054410). Views and opinions expressed are however those of the author only and do not necessarily reflect those of the European Union or the European Research Council Executive Agency. Neither the European Union nor the granting authority can be held responsible for them.

List of acronyms

ACC	Association Cotonnière Coloniale
ACLED	Armed Conflict Locations and Event Data
AFR100	The African Forest Restoration Initiative
ATT	Amadou Toumani Touré
AQMI	*Al-Qaïda au Maghreb Islamique*
BECCS	Bio-Energy with Carbon Capture and Storage
CDM	Clean Development Mechanism
CDR	Carbon Dioxide Removal
CFDT	Compagnie Française pour le Développement des Textiles
CMA	Coordination des Mouvements d'Azawad
CMDT	Compagnie Malienne pour le Développement des Textiles
COP	Conference of the Parties
FAO	Food and Agriculture Organization of the United Nations
FCFA	West African franc
FIS	Front Islamique du Salut
FNL	Front National de Libération
GATIA	Groupe Auto-défense Touareg, Imghad et Allié
GEF	Global Environment Facility
GGW	The Great Green Wall
GHG	Greenhouse Gas
GR	Green Resources
GSPC	Groupe Salafiste pour la Prédication et le Combat
IK	Indigenous Knowledge
IPBES	Intergovernmental Science-Policy Platform on Biodiversity and Ecosystem Services
IPCC	International Panel on Climate Change
ISGS	Islamic State in the Greater Sahara

IUCN	International Union for Conservation of Nature
JNIM	Jama'at Nasr al-Islam wal Muslimin
LED	Low Energy Demand
LK	Local Knowledge
MIT	Massachusetts Institute of Technology
MNLA	Mouvement National de Libération to l'Azawad
MUJAO	Mouvement pour Unicité et le Jihad en Afrique de l'Ouest
NGO	Non-Governmental Organizations
ODEM	Opération de Développement de l'Elevage dans la région de Mopti
OECD	Organisation for Economic Co-operation and Development
ORM	Office Riz Mopti
REDD	Reduced Emissions from Deforestation and forest Degradation
RGPH	Recensement Général de la Population et de l'Habitat du Mali
SEA	Swedish Energy Agency
UCDP	Uppsala Conflict Data Program
UN	United Nations
UNCCD	United Nations Convention to Combat Desertification
UNEP	United Nations Environment Programme
UNFCCC	United Nations Framework Convention on Climate Change
VCS	Voluntary Carbon Standard
WCED	World Commission on Environment and Development

1. Climate security, climate justice and recognition

THE SAHELIAN CRISIS AND THIS BOOK

The droughts and famines of the 1970s and 80s and the Tuareg revolts of the 1990s reproduced a public Western image of the Sahel as a dangerous place at the margins of the modern world. Already during the European 'discoveries' of Africa in the 19th century, the interior of the continent at the fringe of the Sahara Desert was seen as an inaccessible place with its long distance from the sea, its harsh climate, and with the presence of raiding nomadic tribes as well as radical Muslims (or 'jihadists') that did not accept Christians on their land. The myth about Timbuktu as the end of the world epitomised this old European narrative about the Sahel (Benjaminsen and Berge 2004a, 2004b).

More recently, the narrative has been further exacerbated by reports about violent conflicts between farmers and herders, the expansion of jihadist groups, coup d'états, young men leaving to seek work in North Africa or to risk crossing the Mediterranean, and increasing numbers of internally displaced refugees. For more than a decade there has been an escalating security crisis including a spiral of violence that seems to be getting worse every year. This situation has obvious consequences for economies, livelihoods and people's lives in general.

The Sahel has also historically been associated with other tropes, especially related to environmental and climate change. Early in the colonial period, the idea had already emerged among colonial administrators and scientists that the local populations in these drylands were overusing natural resources causing desertification through overgrazing, overcutting of trees and bad farming practices. The solution was seen to be the introduction of European scientific methods and overall colonial governance. This was part of the White Man's Burden.

So, what is this area called the Sahel with so many tropes and myths attached to it? According to Amselle (2022), 'the Sahel' is purely a colonial invention. There is no word in local West African languages for this transition zone between the Sahara Desert and the more humid and forested areas further south. In Arabic 'sahil' means 'coast'. This meaning is also reflected in the word 'Swahili' – a language originating in the coastal areas of East Africa.

After the establishment of the colony French Sudan in 1895, French colonial officers and scientists started to use the term in the early 20th century to refer to 'the coast' of the desert. It emerged for the first time in a report by the French colonial botanist August Chevalier based on a field mission during 1899–1900. The report defined three zones south of the desert according to botanical and climatic criteria – Sahel, Sudan and Guinea. This still reflects a common definition of the Sahel as a climatic zone – for instance, as the area receiving on average between 100 and 600 mm of annual rainfall.

However, politically, 'the Sahel' usually refers to the old French colonies in the dry savanna-dominated landscapes from Senegal and Mauritania in the west to Chad in the east. This is the group of countries that this book refers to as 'the Sahel' (Figure 1.1), although in some definitions, parts of Sudan, Eritrea, Djibouti and Ethiopia are also included in the Sahel.

The current Sahelian crisis is primarily a political and security crisis associated with armed 'jihadist' insurgencies.[1] Jihadist armed resistance against the state in the Sahel first emerged in northern Nigeria in 2009 and northern Mali in 2012, and has later spread to neighbouring areas and countries.

In addition to the security situation, there is an international concern for the consequences of climate change in the Sahel. Global warming affects this region as every other world region. While predictions from climate models vary, most models foresee a slight general increase in total annual rainfall, but

Source: Author's own.

Figure 1.1   Map of the Sahel countries

also with a shorter and more intense and variable rainy season (Biasutti 2013, Giannini 2016). In addition, higher temperatures are observed and expected to continue to increase, combined with potentially more frequent extreme events with heavy storms and long dry spells (Panthou et al. 2014, Guichard et al. 2015).

The increased variability in rainfall is obviously not without serious challenges and risks, especially for farmers, but will not lead to desert-like conditions – contrary to the common image presented not only in the media and by politicians, but also sometimes by activists and even within academic research.

Hence, the most common and mediatised image of climate change in the Sahel persists as a trend toward drier conditions leading to widespread 'desertification'. This public perception continues to dominate despite the fact that the Sahel has become greener since the big droughts in the 1970s and 80s (e.g. Olsson et al. 2005, Hutchinson et al. 2005, Fensholt et al. 2006, Benjaminsen and Hiernaux 2019).

Nonetheless, we keep hearing stories about how serious desertification is in the Sahel. There are two main reasons for why this narrative keeps returning from the dead. First, it is based on an image that easily sticks in one's mind – a creeping desert, swallowing green and fertile areas.

Second, the narrative has been institutionalised, which is a result of the fact that it serves certain interests. This may be foreign governments that have an interest in investing in fighting desertification to obtain a green image and thereby detract attention from the lack of climate or environmental action at home. Or it may be Sahelian governments that use the narrative to attract foreign aid to their country. In addition, a whole industry of NGOs, consultants and scientists emerged from the 1980s that aimed to benefit from the generous flows of funds resulting from the global attention to the desertification narrative.

At first, the process of institutionalisation took place through the establishment of the United Nations Environment Programme (UNEP) in 1972. From early on, the UNEP pointed to desertification as one of the world's most severe environmental problems. Later, it has remained immovable, in contradiction to the international research frontier.

Every year around 17 June, the desertification zombie gets back to life. This is UN's World Day to Combat Desertification and Drought, where we see various initiatives from UN representatives and media coverage from around the world. Still, the UN is not capable of telling us exactly where desertification is taking place. If they for once have a concrete example, the case is often anecdotal and can easily be turned around.

Inspired by the UN and the media publicity, African state leaders have launched the idea to establish a 15km wide and almost 8000km long green wall of trees, going all the way from Senegal to Djibouti, to stop the desert from spreading (chapter 5).

At the 'Conference of the Parties' climate meeting in Paris in December 2015 (COP 21), this gigantic project was promised four billion dollars in donor funding. Besides this it would also get 100 million dollars from the World Bank and additional support from France. In later COPs (Glasgow, Sharm-el-Sheikh, Dubai), more funding has been promised to this project from public and private actors.

In addition to stopping desertification, the goal is to mitigate climate change through tree planting, to reduce migration from the region and to employ young people to stop them from joining jihadist groups. We know, however, from previous tree-planting projects in the Sahel that only a few per cent of the trees that are planted will survive, unless they are watered by hand. We also know that there is not much vacant land in the region, and that introducing new land-use will necessarily dispossess previous land-users and thereby perhaps create more conflicts, or migration (chapter 5).

As already mentioned, the desertification narrative in the Sahel has a long history dating back to early colonial times (Swift 1996, Davis 2016a, Benjaminsen and Hiernaux 2019). In this period, there was a debate among French colonial scientists about desiccation of the dry areas south of the Sahara Desert and whether this was a natural process or caused by local land-use (chapter 2), which resonates with today's popular ideas about climate change and desertification in the Sahel.

During the 20th century, the view that Sahelian farmers and pastoralists create deserts became dominant. This narrative also served to justify colonial state control over land and resources and later served the interests of state-employed foresters, as well as administrators and politicians.

Hence, desertification was thought to be caused by overpopulation, mainly overgrazing and over-cutting of trees, during colonial times and the first decades after independence. But with scientific research demonstrating that the region has actually been greening for the last three decades, this narrative was increasingly questioned in international politics through the 1990s and early 2000s.

However, the idea of 'desertification' is robust and not easy to kill and, with the more recent focus on climate change, the desertification zombie is back again and more alive than ever. In spite of ongoing greening and climate models predicting more rainfall in the region, the dominant media and policy narrative is again one of desertification, but this time the process is mainly believed to be caused by global warming and climatic drier conditions.

This book will, in particular, focus on Mali as a 'hub' country for the current security crisis in the Western Sahel and for debates about climate security and climate justice. These two approaches to understanding people–climate interactions are increasingly represented in public as well as academic fora. The former approach has become prominent in international politics and academically it emanates from International Relations, Political Science and Peace and Conflict Studies, largely with a macro focus on international politics, while often missing out on local and national political, historical and environmental contexts.

Most of the climate justice literature also has such an international or macro focus, but increasingly there are climate justice studies engaging with the more local and detailed aspects of power and justice of climate politics and action. Some of these studies of climate justice are influenced by the critical approach of political ecology and its attempt at bridging local and global levels in the analysis, or studying global processes through the lens of empirical local case studies (Benjaminsen and Svarstad 2021).

The contrast, friction and lack of mutual engagement between climate security and climate justice as two traditions of social science-based environmental and climate research is one of the main themes in this book. In addition, it develops the idea of 'recognition' as a neglected aspect of justice in much of the literature on the Sahel as well as in policy and media debates. The colonial and neo-colonial history in the Sahel demonstrates how Sahelian land-users, including their knowledge and practices, have been consistently misrecognized – both discursively as well as formally (in laws and policies) (Benjaminsen et al. 2022). The book aims to show how a focus on recognition in a historical, political and environmental context may facilitate a deeper understanding of the social processes behind rising insecurity and social-environmental processes more broadly.

Finally, this book combines environmental discourse analysis with a material approach to political ecology. As we shall see, discursive presentations (through rhetorical devices such as narratives and images) of Sahelian populations since early in the colonial period have worked to dispossess people from their land and to put restrictions on their use of natural resources. This means that discourses matter and have material consequences for people's lives.

BACKGROUND TO THE SAHELIAN SECURITY CRISIS

Mali is the key country to study in order to understand the background and causes of the jihadist rebellion and current security crisis in the Western Sahel. Northern Nigeria can be seen to have played a similar role in the Central and Eastern Sahel where the recent wave of jihadist uprising first emerged when Boko Haram turned to violence and launched an insurgency in several northern Nigerian states in 2009 (Comolli 2015, Thurston 2017).

As Boko Haram was taking shape in Nigeria from the early 2000s, Islamist guerrilla fighters from Algeria concurrently moved into northern Mali. This was a spill-over from the civil war in Algeria that had started in 1990 after the Islamist party FIS (Islamic Salvation Front) had won the local elections that year and was set to win general elections in 1992. A military coup, however, put an end to the Islamist hopes of gaining power through elections. Several thousand FIS members were arrested, and the party banned, with the silent endorsement of France and the West in general.

The civil war in Algeria lasted throughout the 1990s until the Islamist guerrilla fighters were either killed or had accepted an amnesty. However, the Salafist organisation Groupe salafiste pour la prédication et le combat (GSPC) did not surrender, and in the early 2000s it extended its presence to northern Mali. After the attacks on 11th September 2001 and the US invasion in Iraq in 2003, the GSPC also changed from an 'Islamo-Nationalist' organisation to one joining a global jihad (Lounnas 2014) and, in 2007, the organisation announced its allegiance to Al-Qaeda and became AQMI (Al-Qaïda au Maghreb Islamique or Al-Qaeda in Islamic Maghreb – AQIM) (Daniel 2012, Harmon 2014, Lounnas 2014).

Since its installation in northern Mali, AQMI managed to strengthen its economic position through the kidnapping of foreigners, receiving millions of Euros in ransom, and through involvement in large-scale smuggling of cigarettes and drugs (Daniel 2012, Harmon 2014, Lounnas 2014). By 2011, this had made it possible for AQMI to build up a military force in northern Mali.

Simultaneously, following NATO's bombing of Libya and the killing of Gadhafi in October 2011, between 1000 and 4000 Tuareg soldiers in the Libyan army had returned heavily armed to Mali (Keenan 2013). This sparked a new Tuareg rebellion that soon managed to occupy the northern towns of Kidal, Gao and Timbuktu. With a lack of resources and low morale, the Malian army soon fled back to Bamako and instead committed a short-lived coup led by Captain Sanogo on 21 March 2012.

In addition to AQMI, the rebel forces consisted of the Salafist groups Ansar Dine[2] and MUJAO (Mouvement pour l'unicité et le jihad en Afrique de l'Ouest),[3] as well as secular Tuaregs fighting for autonomy for northern Mali within MNLA (Mouvement National de Libération de l'Azawad). However, the MNLA became increasingly marginalised by the jihadists because of a lack of resources (Giraud 2013). They also soon became unpopular among parts of the local urban population due to their looting of local businesses in Gao and Timbuktu (Harmon 2014, Bøås 2015). This looting has later been attempted to be justified by the MNLA as revenge for sedentary Songhay supporting attacks by the army and the militia group Ganda Koy on the Tuareg civilian population during the conflict in the early 1990s.

With the marginalisation of the MNLA and a retirement to its stronghold in Kidal, the rebel troops moving south in January 2013 consisted of fighters from AQMI, Ansar Dine and MUJAO. This was seen as a threat to southern Mali and led the Malian Interim President to call Paris for help. On 11th January, 4000 French troops and 1500 vehicles stationed in Chad were dispatched to Mali in what was named 'Operation Serval' (Hanne 2014). At the village of Konna in the Mopti region, the French forces managed to push back the rebel troops and advance to take back control over the towns in northern Mali.

Jihadist groups have, however, gradually moved back into the Mopti region since 2015, but not through a large-scale collective invasion. This take-over first happened through piecemeal infiltrations by smaller armed groups, which have stayed 'in the bush' and occasionally visited villages to give their orders or to assassinate people who they have seen as collaborators with the government or the army. Locally they are talked about as 'les gens de la brousse' (the people in the bush).

After the Serval operation, France established Barkhane in 2014 as a permanent force in the Sahel with 4800 soldiers operating in five countries from Mauritania to Chad. In Mali, Barkhane had about 1000 troops stationed in Gao – until August 2022 when the French troops pulled out after a disagreement with the new military government in Mali that had taken power in a coup two years earlier.

This disagreement reflects general and growing anti-French and anti-Western feelings in the Sahel, which are found among urban youths in particular, but they are also reflected in the media, not least in social media, and shared by many intellectuals in the region.

However, just after Serval and French troops freed the northern cities of Timbuktu and Gao in 2013, people were waving the 'tricolor' welcoming the French army, and former President Hollande came on a triumphant visit to Timbuktu where he was celebrated as a hero. But the public view on French intervention thereafter changed rather quickly, although there has always been scepticism of French political and economic interests within sections of the Sahelian population.

Since then, a lot has gone wrong in the relationship between France and Mali, but also between the former colonial power and several other Sahelian countries. Different jihadist groups have continued to expand their area of control. In addition, young officers with an anti-French and pro-Russian rhetoric have taken power from democratically elected governments in Mali (August 2020) and in Burkina Faso (January 2022). In addition, large popular demonstrations in Bamako and Ouagadougou have supported the coup-makers.

In July 2023 there was also a coup in Niger – this time apparently not driven by anti-French attitudes, but by an internal power struggle among political and military elites. Popular demonstrations in the capital Niamey, however, supported the coup using the same rhetoric as in the neighbouring countries. In this way, the coup in Niger has also been seen as part of a new anti-French wave.

Since independence in the early 1960s, there has been both a close and tense relationship between France and its old colonies in the Sahel. The Sahelian elite knows France and the French language very well. They may send their children to schools where they learn more about France than about African society and history.

Still, anti-French, or anti-colonial attitudes are not new. They have been there since colonial time – from the anti-colonial struggle of the Islamist Samory Touré and the Tuareg leaders Kaocen and Firhoun in the early 20th century to Thomas Sankara's short presidential period in Burkina Faso until he was murdered in 1987. Sankara has later been glorified, especially by young people in the Sahel, as an African Che Guevara, because he stood up against France and the West and because he fought against the widespread corruption among the political and administrative elites.

The French colonisation was in many places brutal. If there was resistance, villages could be bombarded and all the inhabitants killed (Sûret-Canale 1962, Lefebvre 2021). In addition, while the Second Republic formally abolished slavery in the colonies in 1848, this was never applied in Senegal. In 1855, this practice was also formally expressed by the administrative council of the colony. During the colonial conquest, captives were distributed to officers and soldiers and were also used as recruits in the army (Sûret-Canale 1962).

Later, so-called 'freedom villages' were established for slaves freed from internal control in African societies (e.g. Bamabara, Fulani, Tuareg), which often functioned more as labour reserves and were talked about as 'the villages of the Commander's slaves' (Sûret-Canale 1962). Inhabitants in these 'freedom villages' were used to supply labour for the colonial government, for missions, and as wives for soldiers, and they needed special permits to move away. Fugitives were punished with prison.

The various forms of extraction of labour also clearly impeded local development in the Sahel. As Sûret-Canale (1962: 68) puts it:

> by systematically extracting the fruits of surplus labour and even a large part of essential labour, (colonialism) prevented any form of local accumulation of capital among the Africans. The extreme misery of the standard of living formed an obstacle to any technical improvements. … Any possibility of developing indigenous capitalism, or an indigenous bourgeoisie was thus extremely restricted.

In addition, soldiers from the Sahel also fought in the French army during both the First and Second World Wars as well as in the war in 'Indochine' in the 1950s. Several hundred thousand African soldiers lost their lives during these wars and, after the war effort, those who survived were generally sent back home without much compensation. This shameful treatment of 'les tirailleurs sénégalais' has only recently been openly discussed in France.

In the 1940s, a special French version of neo-colonialism started to emerge, which has been called 'Francafrique' (Borrel et al. 2021). The model builds on personal friendships among French and African political elites. Gifts and services are exchanged, for instance between French politicians and African presidents and ministers as part of strategic alliances where official acts and personal relations are mixed together. In other words, this is high-level corruption.

Curiously, both the recent coups in the Sahel and their popular support as well as the rise and growth of jihadist groups can be seen as reactions to Francafrique. In fact, the new military regimes as well as the jihadists, who are fighting each other, both have an anti-elite and anti-corruption rhetoric and aim to change the old neo-colonial system and untie the links to France.

The anti-French feelings in Mali have grown because many people do not understand how almost 5000 well-equipped foreign troops have not been able to prevail over the jihadists who have much less sophisticated weapons. This has led to theories of conspiracy about France not wanting to end the insecurity in the Sahel, because it leads to weak states that can be more easily exploited, which is not an uncommon view in the Sahel or among the Sahelian diaspora. This view is also reproduced by a few popular bloggers on social media with several hundred thousand followers and who also have a clear pro-Russian agenda. In this narrative, Russia is presented as a global power with an anti-colonial agenda.

Obviously, it has not helped the popularity of France in the West African francophone countries that former President Sarkozy in a famous speech at the Cheikh Anta Diop University in Dakar in 2007 said:

> The tragedy of Africa is that the African man has not entered history enough .... He never rushes towards the future .... In this universe where nature commands everything, man remains motionless in the midst of an immutable order where everything is written in advance. ... There is no place for human adventure, nor for the idea of progress.

This speech, stating that Africans are trapped in superstition and traditions and not interested in progress, sparked widespread responses throughout Francophone Africa. It also bolstered a public image of French neo-colonial views on Africa. Likewise, President Macron in a speech at the Elysée Palace in August 2023 said, referring to French military intervention in the Sahel, that:

> If France had not acted, if our troops had not been killed in action in Africa, if Serval and then Barkhane had not been set up, today we would no longer be talking about Mali, Burkina Faso or Niger. These States would no longer exist with their current borders. I can say that with certainty.

The message here is that the Sahelian countries exist because of France, which caused an outcry on social media in particular, and spurred further anti-French sentiments.

In Mali, the French forces became unpopular, in particular after they left the Kidal area in the north to the Tuareg rebels instead of to the Malian army. This reinforced a common view that France supports the independence struggle of the Tuareg. In addition, since Ansar Dine seems to hold considerable power in the Kidal area, including collecting a 10 per cent tax (*zakat*) on the artisanal gold production in the area, this act was also seen by many in Mali as evidence that the French are not interested, for strategic purposes, in fighting the jihadists.

When the French forces started to plan their withdrawal from Mali, the new military government invited in the infamous Wagner group to replace them. This group has more recently, after the death of its leader Prigogine, been renamed the Africa Corps, which is a chilling reference to Nazi Germany's 'Afrika Korps' that fought in North Africa from 1941 to 1943 under the command of General Rommel.

For those who had already followed Wagner's warfare in Ukraine, Syria and other countries, this intervention in Mali has not surprisingly led to serious human rights violations. The most mediatised event happened in the village of Moura in central Mali in March 2022 when more than 500 people were massacred. In May 2023, the Office of the High Commissioner for Human Rights published a report on the massacre, which was rejected as French and Western propaganda by the military government. The official story was that the Malian military had run an operation that resulted in the killing of 203 terrorists.

The French withdrawal from the camp in Gao has also opened up space in the northeast for the Islamic State in the Greater Sahara (ISGS) to expand its presence in the area called the Liptako-Gourma. ISGD is a Fulani-dominated group centred around this tri-border area between Mali, Niger and Burkina Faso.

In addition to these jihadist groups, there are currently the 'secular' Tuareg rebel groups under the umbrella of CMA (Coordination des Mouvements d'Azawad) and government allied militia such as the Tuareg-based GATIA (Groupe Auto-défense Touareg, Imghad et Allié), the Fulani-based Ganda Iso, the Songhay-based Ganda Koy and the Dogon-based Da Na Amassagou. This range of armed groups is, in addition to the Malian army, responsible for an increasing level of violence. According to the Uppsala Conflict Data Program (UCDP), the number of battle deaths in Mali increased, for instance from 182 in 2016 to 3641 in 2022.

There is, however, a certain plasticity when it comes to the organisation and membership of the armed groups in northern and central Mali. Since 'alliances form and collapse at a high rate between a myriad of armed groups, which, to complicate matters, may rebrand themselves under new acronyms' (Desgrais et al. 2018: 656). Individuals may also change allegiance from one group to another. Therefore, 'membership' in the various groups as well as their external discursive politics, should be treated with caution.

For instance, jihadist, secular and government-allied Tuareg groups have in 2023 been seen to cooperate in a joint, but unsuccessful, attempt to fight ISGS in the Menaka area. This Tuareg alliance seems to initially have responded to a longer history of conflicts over pastures and seasonal lakes between Fulani and Tuareg in the Liptako-Gourma. Alliances have, however, since broken up again into more complicated patterns with some Tuareg groups also joining the Fulani-dominated ISGS. These shifting coalitions make security analysis challenging and with a short best before-date.

CLIMATE SECURITY

In international politics, climate change has during the last couple of decades increasingly been framed as a security issue, which in the academic literature is referred to as the 'securitization of climate change' (Brown et al. 2007), or simply as 'climate security'.

This international attention to climate security is found in particular among policy, military and NGO actors, while relatively few researchers advocate this idea (Selby and Hoffmann 2014). 'In this, the climate security field diverges sharply from other areas of climate change policy, where scientists have played formative … roles in pushing forward national and international action' (Selby and Hoffmann 2014: 749).

Furthermore, since climate security has continued to gain traction during the last decade, there seems to be an increasing number of applied and policy-orientated researchers with modest empirical ambition who align themselves with this approach, mainly following pressure from policy-making combined with financial opportunities.

Popular media is also ripe with stories of not only climate disaster in African drylands, often leading to desert-like conditions, but also connecting such events to widespread violence. Deserts and violence are closely connected in the public image.

For instance, in the dystopic movie series 'Mad Max' an environmental disaster, presumably caused by climate change, has created a desert-like planet dominated by the violent struggle for survival over scarce resources among the remaining few people on Earth. This scenario also resonates with some popular books on 'climate wars' (e.g., Dyer 2010, Welzer 2012).

According to Dyer (2010), an 'apocalyptic crisis' is 'set to occupy most of the twenty-first century' with 'a probability of wars, including even nuclear wars, if temperatures rise two to three degrees Celsius' (Dyer 2010, xi–xii).

In Welzer's book, however, climate change is not seen as the direct cause of conflicts, but more as one of the underlying factors creating declining food production, increased land degradation through more droughts and floods, increased health risks, and more people on the move. Hence, 'the consequences of climate change will reinforce and deepen survival problems and the potential for violence; they will interact with political, economic, ethnic and other social-historical factors and may also lead to open use of force' (Welzer 2012: 75). This thinking is in line with numerous other security analysts' writings and is frequently referred to as a 'threat multiplier', resulting from a collision of political, economic, and environmental disasters (Parenti 2011).

In a historical perspective, this new 'climate reductionism' (giving climate the role of a key variable predicting social change) may be compared to the 'climate determinism' of the nineteenth and early twentieth century (Hulme 2014). As in the heyday of climate reductionism when the agency of colonial subjects was reduced to being a product of African climates, there is also a particular focus on Africa in today's climate reductionism. African drylands, and in particular the Sahel, are usually pointed out as the most prominent example where there is a volatile mix of climate change, drought, food shortages, migration, and armed insurrection. The Sahel was also used by the Norwegian Nobel Committee as the prime example of the link between climate change and conflict when former US Vice President Al Gore and the IPCC were awarded the Nobel Peace Prize in 2007 (see Box 1.1 below with an example of how Artificial Intelligence (AI) sees climate change and conflicts in the Sahel).

BOX 1.1   CLIMATE CHANGE AND CONFLICT IN THE AFRICAN SAHEL: A DESCRIPTIVE OVERVIEW (AS SEEN BY AI)

The African Sahel, a semi-arid region stretching across several countries in North and West Africa, has been significantly impacted by both climate change and violent conflict. Here's a detailed description:
1.   Climate Change Effects
   •   Increased Temperatures: Rising temperatures in the Sahel exacerbate existing challenges related to water scarcity, food security and livelihoods. Prolonged droughts and heatwaves lead to crop failures, livestock losses and reduced access to clean water.
   •   Erratic Rainfall Patterns: The Sahel experiences irregular rainfall patterns due to climate change. Unpredictable rainy seasons affect agricultural productivity, leading to resource scarcity and competition.

- Desertification: The expansion of desert areas into previously habitable regions threatens communities' survival. Soil degradation and loss of vegetation contribute to desertification.
- Migration and Displacement: Climate-induced environmental stress drives migration within and across Sahelian countries. Displaced populations often face resource conflicts and tensions with host communities.

2. Conflict Dynamics
   - Resource Scarcity: Scarce resources, including water, pasture, and arable land, create competition among pastoralists, farmers, and other groups. This competition can escalate into violent clashes.
   - Ethnic and Communal Tensions: Ethnic divisions and historical grievances intersect with resource scarcity. Conflicts between ethnic groups over land, cattle and access to resources are common.
   - Extremist Insurgencies: The Sahel is a hotspot for extremist groups like Boko Haram, Al-Qaeda in the Islamic Maghreb (AQIM), and the Islamic State in the Greater Sahara (ISGS). These groups exploit local grievances, weak governance and poverty to recruit fighters and destabilise the region.
   - State Fragility: Weak governance, corruption, and inadequate state institutions contribute to insecurity. Armed groups exploit governance gaps and lack of effective law enforcement.

3. Interconnected Challenges
   - Feedback Loop: Climate change exacerbates existing vulnerabilities, leading to conflict. In turn, conflict disrupts development efforts, exacerbating environmental degradation and resource scarcity.
   - Humanitarian Crisis: The combination of climate-induced displacement, conflict and food insecurity creates a complex humanitarian crisis. Displaced populations face violence, hunger and inadequate access to basic services.
   - Regional Cooperation: Addressing climate change and conflict requires regional cooperation among Sahelian countries. Collaborative efforts are essential to mitigate environmental stress and prevent further conflict.

In summary, the African Sahel faces a precarious situation where climate change intensifies existing challenges, leading to violent conflict.

Sustainable solutions must address both environmental resilience and peacebuilding to ensure the well-being of Sahelian communities.
Source: Copilot, prompt: 'Give a description of climate change and conflict in the African Sahel'.

Curiously, the climate-conflict narrative as used on the Sahel (and reflected in Box 1.1) also includes Malthusian elements (Hartmann 2014, Selby and Hoffmann 2014) in its focus on resource scarcity as a cause of environmental degradation, poverty, and an escalating struggle for resources, and with roots in the 'environmental security' school that emerged in the 1990s (Homer-Dixon 1994). Traditionally, this literature has primarily been concerned with 'overpopulation' and the associated 'overuse' of renewable natural resources as causes of conflicts (Homer-Dixon 1999), but with global warming and the subsequent climate security narrative, the anticipated impact of anthropogenic climate change on the security of societies and livelihoods has gained prominence.

It is, however, important to stress here that questioning the climate-conflict narrative in the Sahel does not in any way mean that climate change should not be taken seriously or that it does not represent serious risks affecting the region. The problem is rather that the narrative depoliticises and dehistoricizes conflicts and reproduces external and rather colonial views of the Sahel. This implies that Sahelian historical and political contexts are not recognised in this narrative. It seems easier and more convenient for various actors – e.g. some governments, activists and even some academics – to stress climate change as a driver rather than trying to understand a complex context.

Even though climate change is obviously affecting Ukraine as well as Israel/Palestine, it would be ridiculous to allude to these violent conflicts as in any way climate-driven, because the histories and political contexts of these conflicts are well known and are repeatedly presented by leading media. An interest in context is, however, largely absent when it comes to conflict 'analysis' in the Sahel.

However, questioning some of the climate security literature and associated claims clearly involves a dilemma because anthropogenic climate change and the need for urgent climate action deserve all the attention they can get. Claims made to draw attention to this global problem need, however, to be based on facts. If not, there is also a risk of undermining public engagement in the issue in the long term.

In the climate security framing, 'security' and 'insecurity' are usually understood in relation to the risks of violence. The general idea is that climate change plays a role in one way or another in producing a violent conflict.

Some climate security analysts do, however, define security more broadly. For instance, Busby (2022) includes humanitarian emergencies as an outcome of climate change in the concept of 'security' together with violent conflict. He criticises the climate security literature for not paying enough attention to causal mechanisms. By a comparative case-study analysis of seven country cases in Africa and Asia he concludes that there are three factors in particular that tend to lead climate change to trigger conflict and humanitarian emergencies. These are 1) weak state capacity; 2) exclusive political institutions; and 3) foreign assistance that is blocked or delivered unevenly. The final chapter of this book comes back to a critical discussion of these factors.

Moreover, McDonald (2021: 4) identified four different climate security discourses reflecting 'whose security is prioritized, from what threats, by what agents and by what means'. These discourses focus respectively on the nation-state, international society, people or ecosystems. Adaptations to climate risks will vary with these various discourses.

McDonald's own approach lies within the last of these four discourses seeing climate security within a broader environmental philosophy framework, and extending the attention from the climate system to the resilience of ecosystems. It is, however, hard to see what this means in practice. What type of action may follow from such an ecological security discourse?

While McDonald (2021) acknowledges that climate action may hurt vulnerable people, his focus is still on progressive and effective climate action challenging 'the tyranny of the status quo' (p. 192) without much attention to the trade-offs and hard choices involved. One wonders what this would mean in terms of concrete climate action in the Sahel (Benjaminsen 2023), because the trade-offs are numerous related to the security risks of climate mitigation leading to forms of green neo-colonialism (or green grabbing – Fairhead et al. 2012) (see chapter 5).

The approach of the IPCC to climate security has changed with the last few Assessment Reports. Referring to the Fourth Assessment Report, Nordås and Gleditsch (2007: 628) stated that 'even the IPCC, which rightly prides itself on being a synthesis of the best peer-reviewed science, has fallen prey to relying on second- or third-hand information with little empirical backing when commenting on the implications of climate change for conflict'. With the publication of the Fifth Assessment Report in 2014 there was, however, considerable improvement in how human security was discussed, but still with some inconsistencies between the various chapters depending on varying authorship (Gleditsch and Nordås 2014).

While the Sixth Assessment Report did not have a separate chapter discussing human security, the report focused on how climate change-related risks may occur via possible indirect pathways, such as for instance through reduced food security or the implementation of adaptation or mitigation measures

(O'Neill et al. 2022). In fact, recent research highlights state capacity and the level of socio-economic development as generally more important factors than climate change in explaining conflict, while also stressing that climate change-related risks may potentially impact conflicts (Mach et al. 2019). Increasingly, the conflict risks associated with climate mitigation initiatives such as afforestation, windmills, and biofuels are also discussed in this recent literature on climate change-related risks (Buhaug et al. 2023).

This means that there clearly is a link between climate change and conflict, but what that link implies, and the understanding of this link vary substantially among researchers, journalists, climate activists and policy-makers. Such a potential for misunderstanding is also mobilised by actors with an interest in pushing the climate-conflict agenda stating, for instance that 'only a "misreading of the state of science" could allow any doubt over the links between climate change and insecurity' and, hence, the UN Security Council should act on the science and put climate security challenges on its agenda (Day and Krampe 2023). The risks of such a focus are that the basic causes of conflicts are glossed over or simply neglected, and that possible pathways to solving conflicts politically are not sufficiently explored. Hence, climate change, while a dangerous on-going process, may serve as a political lightning rod inhibiting political and media attention to genuine causes behind conflicts.

In addition, what exactly should the UN Security Council do 'to act on the science' in a region such as the Sahel – build another wall of trees (chapter 5) or initiate a military intervention to stop migrants from moving towards Europe?

In addition to these risks inherent in the climate security frame, this frame may also serve as an 'anti-politics tool' for regimes that are politically responsible for conflicts. For instance, the former Malian President Amadou Toumani Touré (2002–2012), when meeting foreign leaders and audiences, often stressed climate change and desertification to explain the security crisis in the northern part of the country. In a similar vein, President Assad of Syria blamed the civil war in the country on drought and global warming when meeting with diplomats, while former President Omar al-Bashir in Sudan liked to give a similar explanation of the Darfur conflict (Selby et al. 2022).

International media have also generally been keen to repeat a policy narrative about climate-caused conflicts. Especially the current crisis in the Sahel has drawn international attention to climate change as a cause. Just to give two examples – Le Monde reported on 11th April 2019 that conflicts between Fulani herders and Dogon farmers in Mali are caused by resource scarcity following climate change and population growth, while Deutsche Welle (11th June 2019) concluded that 'The conflict between Dogon and Fulani ethnic groups over resources in Mali has been exacerbated by climate change, population growth, an absentee state and Islamism' (see chapter 4 for a discussion of Fulani-Dogon conflicts in context).

Such policy- and media-driven statements stand in contrast to most research in the field. Quantitative peace and conflict studies have questioned general assumptions about climate change as a driver of violence and insecurity, although there may be indirect pathways under certain contexts (e.g. Buhaug 2010, Theisen et al. 2013, Koubi 2019, Buhaug et al. 2023), while case-based political ecology research in the Sahel has pointed at the flaws in the assumed links between climate and conflict highlighting political, historical and social contexts to explain conflicts (Benjaminsen et al. 2012, Abrahams and Carr 2017, Selby et al. 2022)

CLIMATE JUSTICE AND RECOGNITION

The field of climate justice started to take off from the early 2000s with a growing realisation that climate change as well as its responses through mitigation and adaptation tend to produce inequalities (Jafry et al. 2018). Especially after COP15 in Copenhagen in 2009, a global movement of climate justice was established as a response to what was widely acknowledged as a failed meeting (Schlosberg and Collins 2014). While the term is understood differently by climate activists, academics and policy-makers, a commonality is a concern for fairness, equity and justice in the context of climate change (Jafry et al. 2018).

Climate justice has become well established in the international politics of climate change including the UNFCCC and the Paris Agreement. Also the IPCC, through its various reports, has reflected an increased focus on climate justice with it being one of the main themes in the Sixth Assessment Report (Ara Begum et al. 2022).

National governments, especially from the Global South, as well as many activists tends to focus on the distributive injustice of climate change impacts combined with the historical responsibilities of the Global North, which is recognised in the UNFCCC through the principle 'common but differentiated responsibilities' (Fisher 2015).

In its Sixth Assessment Report, the IPCC also argues that there is increasing global realisation that the urgency of the climate crisis necessitates swift action to initiate a green and societal transformation to a low-carbon or green economy. Such a transformation would necessarily imply deep structural societal change that will have implications for social justice (Ara Begum et al. 2022).

To pursue this aim, there are many potential sustainabilities to choose from (Cavanagh and Benjaminsen 2017). These range from more 'business-as-usual' scenarios focused on modernisation, technological change and marked-based solutions within what has been termed 'ecomodernism' (Asafu-Adjaye et al. 2015) to more radical propositions such as 'degrowth' focused on down-scaling production and consumption in the Global North (D'Alisa et al. 2014). While there may be many potential trade-offs between justice and efficient climate action, for social transformation to succeed and be sustainable in the long term, it would need to build on popular understandings of what is fair and socially acceptable (Scoones et al. 2015).

With the Rio+20 conference in 2012, 'green economy' emerged as a new tool to achieve sustainable development (UNEP 2011, OECD 2012, World Bank 2012). However, leading approaches to the green economy differ in the global North and South (Bergius and Buseth 2019). In the North, its main component has been dominated by technological and market-based solutions in the renewable energy sector. This is also important in the global South, but in practice, green economy implementation in the South has tended to focus on environmental protection initiatives combined with shifts in access and control over renewable natural resources with adverse social implications (Cavanagh and Benjaminsen 2017, Bergius et al. 2018). Hence, on-going policy-led transformations towards a green economy take different directions in the Global North and South, and will also create different winners and losers reflecting their access to power.

These debates broaden up the question of climate justice beyond the question of distribution among nation-states. At the national and local level, important questions with justice implications are not only where the effects of climate change will be felt, but also – Who will bear the costs of mitigation? How will decisions about adaptation and mitigation be made? Whose interests and livelihoods count in mitigation and adaptation policies? (Fisher 2015).

In academic work, climate justice may be seen as a sub-field of environmental justice, with its dominant framework first developed by Schlosberg (2003). This framework highlights three elements of justice in particular – distributive justice, recognition and procedural justice – and has again taken inspiration from political philosophy, in particular the work of Nancy Fraser (1998), but also that of Iris Young (1990) and Axel Honneth (1995, 2001). Nancy Fraser initially proposed a two-dimensional focus on distribution and recognition, but in the second half of the 1990s she included representation (or procedural justice) as a third dimension (Fraser 1998).

In discussions about climate justice, distributive justice refers to the distribution of burdens and benefits related to climate change itself or to climate action. Justice as recognition concerns who is given respect (or not) and whose interests, values and views are recognised and taken into account. Procedural justice is about who is involved and has influence in terms of decision-making.

Moreover, Fraser (2000) connects recognition to social status and sees misrecognition as the institutionalisation of social subordination. Such misrecognition may take place in different ways, for example, through cultural domination, non-recognition (or lack of recognition) or disrespect. Misrecognition may be connected to social categories such as gender, race, religion or ethnicity.

Fraser sees recognition and distribution as two parallel dimensions of justice. This is in disagreement with Honneth who argues that recognition is the fundamental and overarching category of justice, which means that questions of distribution and redistribution are derived from recognition (Fraser and Honneth 2003). However, to Fraser, 'not all maldistribution is a by-product of misrecognition' (Fraser and Honneth 2003: 35). The injustice and maldistribution produced by speculative capitalism may not necessarily be linked to misrecognition, she argues.[4]

One may, however, argue that without some form of recognition, it is unlikely that a group of people will benefit from distributive or procedural justice, since recognition concerns who is given respect (or not) and whose interests, values and views are recognised and taken into account. In the context of climate change, recognition may be seen as referring in particular to whose knowledge, interests, priorities and livelihoods that are considered valuable in social constructions such as leading discourses and narratives, as well as in politics and practice.

While Fraser tends to focus on legal or formal recognition primarily through state institutions, Honneth conceives recognition as containing two dimensions – legal recognition in terms of formal rights and intersubjective recognition consisting of solidarity and love (J. Fraser 2018). In climate discourse, policy and practice, misrecognition may be linked to both a lack of formal rights as well as a lack of solidarity with marginalised peoples among powerful actors and policymakers.

In discussing misrecognition, Fraser (2009) develops further Hannah Arendt's concept of 'misframing'. This refers to who has the right to have rights and the fact that some people become non-persons with respect to justice. While those who suffer may become objects of charity, they remain without any formal rights to justice. These misframings may shield powerful states and transnational companies from the reach of justice, and it may be seen as the defining injustice of a globalising age, according to Fraser (2009). Such politics of framing, which refer to who counts as an object of justice, make invisible both subaltern groups who suffer from injustice as well as the power of countries, institutions or companies who are at the source of this injustice.

But, while both Arendt and Fraser focus on formal justice, the notion of misframing can be extended also to intersubjective justice. Marginalised people are made invisible not only as legal objects of justice, but also in leading discourses and practices as actors in their own right. Their interests, priorities and livelihoods are neglected. They are therefore not only outside the realm of formal justice, but are also neglected and misrecognized discursively, which reflects a non-formal type of misrecognition that may be seen as lack of solidarity. Hence, this book discusses not only formal (mis)recognition, but also intersubjective forms expressed through discursive (mis)recognition.

Such lack of solidarity may be associated with a lack of interest in the historical and geographical context of the injustice taking place. While such disinterest in context will be partly produced by disciplinary backgrounds, it also has deep historical roots in colonialism and inequalities in voice and perspective that have remained until today.

When it comes to issues of climate justice in the Sahel, the perspectives are diverse. On the one hand, there is an NGO and climate activist narrative saying that the Sahel is a region that is particularly vulnerable to climate change and that this climate-induced vulnerability is also is an issue of climate justice. The argument is that the vulnerability to climate change in the Sahel comes from rising temperatures and more rainfall variability. The expected continued increase in total rainfall and the greening of the Sahel over the last decades are rarely mentioned.

Instead, the desertification narrative lives on. Since the Sahara Desert is spreading south due to emissions of greenhouse gases, the argument goes, it is a climate justice duty of rich nations to fund initiatives to stop this process – through projects such as the Great Green Wall (see chapter 5). Hence, climate finance for mitigation and adaptation as well as for loss and damage remain the focus of this narrative reflected in numerous NGO websites.

In contrast to such an apolitical narrative, Kashwan and Ribot (2021: 326) find that such an agenda

> (diverts) attention from the inequalities that make people vulnerable and make climate events – whether ordinary or intensified by global change – dangerous in the first place (and that) governments in the global North and South alike are now blaming the climate and weather for crises that stem from inequality – avoiding blame for conditions that they created and could redress.

Hence, vulnerability to climate change has political and historical causes linked to exploitation and marginalisation and should not primarily be attributed back to climate change. Vulnerability does not fall from the sky, as Ribot (2010) put it.

Related to this political-economic critique of apolitical views of vulnerability, 'agrarian climate justice' has emerged from critical agrarian studies as an alternative view on climate justice in the rural South, including in the Sahel (Borras and Franco 2018). This view is focused on:

> concerns for land rights, people's sovereignty over natural resources and the importance of sustainable farming, as seen in landmark documents of climate justice movements, such as the 2007 Bali Principles of Climate Justice and the 2010 People's Agreement of Cochabamba (Calmon et al. 2021: 2796).

This perspective combines concerns for agrarian justice and climate justice and builds on five normative goals: Redistribution (of access and control of land and natural resources), recognition (of local rights to land and resources), restitution (of lost access to land), regeneration of land (ecologically and economically), and these four goals can only be achieved through resistance within or against capitalism (Borras and Franco 2018: 1319).

There is a long way from this radical perspective on climate justice to mainstream views among many NGOs and climate activists linking climate justice to the lack of climate finance for interventions in the Global South. The dominant perspective tends to be that there is a funding gap compared to the needs in the South as a result of unkept promises from rich countries, for instance through pledges made at the COP in Paris in 2015. The quality of projects and the risks of new projects reinforcing climate injustice are rarely discussed. This position is also reflected in the chapter on climate finance in the Sixth Assessment Report of the IPCC, which recommends accelerated finance for nature-based solutions including mitigation in the forest sector (afforestation and REDD), without discussing the potential risks of climate injustice of such interventions (Kreibiehl et al. 2022).

THE REMAINDER OF THIS BOOK

The next chapter goes through the history of the desertification narrative in the Sahel since it emerged early in the colonial time. It also discusses the impact of this narrative on the livelihoods of people in the Sahel. This impact was in particular felt through the establishment of the militarised colonial Forest Service that was continued as a 'forest police' after independence and further strengthened by international aid during the rise of the sustainable development agenda in the 1980s. In this way, the chapter provides a background to discussions of both climate security and climate justice in the Sahel.

Chapter 3 discusses pastoralism as livelihood strategy in the Sahel. Furthermore, it takes inspiration from the analytical lenses of 'moral economy', 'elite capture' and 'land dispossession' to help provide a context to explain why so many Sahelian pastoralists have joined jihadist groups. The chapter also highlights the role of the two jihadist leaders – Iyad Ag Ghaly and Hamadoun Kouffa – in establishing the uprising in Mali. The overall aim of the chapter is to demonstrate how context matters in attempts to understand conflict and revolt within the broader issue of climate security.

The importance of understanding context related to climate security is also the main message in chapter 4. This chapter discusses recent conflicts between Dogon and Fulani on the Seeno plains in central Mali. These conflicts have been presented as typical farmer–herder conflicts driven by scarcity of land and natural resources, and sometimes associated with climate change. In contrast to such a scarcity narrative, the chapter argues that these conflicts may rather be understood as an insurgency and its counter-insurgency. Many Fulani have joined jihadist groups in the area, and when the Malian army was unsuccessful fighting these groups, it supported a Dogon militia that attacked Fulani villages, which again led to counter-attacks from Fulani/'jihadi' groups. In addition to demonstrating the complexity of the current crisis in Mali and how simplistic narratives of climate- or scarcity-driven conflicts are misleading, the chapter also shows how views of the enemy as 'terrorists' or 'jihadists' are dangerous and able to further fuel violent conflicts.

Chapter 5 first discusses ecomodernism versus degrowth as two contrasting approaches to green transformation with their different implications for climate justice. Ecomodernism is the dominant approach to transformation to a low-carbon economy in rich nations as well as in international organisations including the IPCC. It is, however, built on the flawed assumption of large land areas in the Global South, especially in Africa, that are unused and without any ownership, and that therefore are available for large-scale afforestation. Ongoing afforestation projects, such as the plantations of Green Resources in East Africa and the Great Green Wall of Africa in the Sahel, have demonstrated the climate injustice of such projects as they lead to land dispossession and potentially also to resistance and land-use conflicts. When there is such straightforward misrecognition of the local population, there is a clear risk that climate mitigation through afforestation may lead not only to a failed climate project, but also to increased resistance, which might again exacerbate conflicts. In addition, planting trees on grasslands and savannas may not be an efficient way to store carbon and it may also have adverse effects on biodiversity. Despite these obvious problems, afforestation is hyped as one of the main approaches to mitigate climate change globally. This is possible because of carbon fetishism meaning that the adverse effects of climate mitigation are not visible to the general public. The gap between how a project is sold in websites, project documents and social media and what happens on the ground becomes bigger the more global the commodity is and the longer the distance is between production and consumption. Such lack of knowledge about context may also 'take revenge' on projects. What is often presented as unexpected consequences could have been predicted if context were considered. Avoiding such adverse consequences requires taking context and the insights from qualitative research seriously. It also requires recognising the knowledge of local people and their history of land use when climate projects are planned.

Chapter 6 is the concluding chapter. It sums up the book's approach to climate security and climate justice, which are two traditions of social science-based environmental and climate research that are rarely brought in dialogue. In particular, the book aims to fill the gap in the literature in bringing discussions of climate security in dialogue and confrontation with climate justice.

This book is further inspired by discussions about 'recognition' as a key aspect of justice in addition to neo-Marxist approaches within peasant and critical agrarian studies explaining the emergence of peasant rebellions. In addition, it draws on environmental discourse analysis demonstrating how discursive presentations of Sahelian societies and landscapes since early in the colonial period until the present have worked to dispossess people from their land and to put restrictions on their use of natural resources. Finally, this concluding chapter argues that a good comprehension of context related to climate security and climate justice depends on qualitative research to understand social dynamics, power relations and historical trajectories. But it also depends on sound natural science to assess landscape and ecological change over time as well as past and future climate change. There is, however, a certain robustness and resilience of Eurocentric and neo-colonial narratives about the Sahel, which tend to prevail over scientific facts and knowledge of context. These narratives tend to be driven by political interests in sustaining ecomodernist approaches to climate and environmental governance in the rich countries of the North.

NOTES

1. Jihadism is not an unproblematic term. Jihad traditionally refers to a moral or spiritual rather than a violent struggle (Sedgwick 2015). However, here the term is used to refer to insurgency groups in the Sahel that refer to themselves as 'jihadist' or 'mujahideen'.
2. This is a Tuareg-based jihadist organization led by Iyad Ag Ghaly originating in Kidal and demanding that Mali becomes an Islamic state.
3. This was a rather vague Salafist organisation with unclear leadership, but with a base mostly among Songhay and Fulani and with an origin in AQMI.
4. For instance, Fraser gives the example of how white male industrial workers may be unemployed through a factory closing because of a speculative corporate merger (Fraser and Honneth 2003: 35).

2. Colonisation, desertification and forest governance

SCIENCE AND POLITICS (NEVER SHALL THE TWO MEET)

There are few issues in society where the gap between science and politics is wider than on the issue of desertification. Such a gap is often pointed at in relation to climate change, but this is a different sort of gap where it is not mainly the analysis of the issue itself, which is the problem, but rather the extent and nature of action.

On desertification, especially on the particular issue of desertification in the Sahel, the world region most often referred to as the worst hit by this problem, scientists have for 30 years or more amassed evidence against the idea of a Sahara Desert crawling southwards. Still, however, the UN, the World Bank, major donors of development aid, powerful politicians, big and small NGOs, and even some climate activists continue to focus on this imagined problem. The media is no better. One often wonders where critical journalism has gone.

As Rudyard Kipling famously said in the poem 'The Ballad of East and West' (1889), 'never the twain shall meet', this also seems to be the case on the question of science and politics related to desertification in the Sahel. While scientific research published during the last three-four decades has demonstrated a greening of the Sahel, 'desertification' has simultaneously reached the status of a global environmental issue. It has become an institutionalised fact. But before we discuss the reasons behind this lack of 'meeting', let us take a look at the history of this environmental issue.

At the first global conference on the environment, the United Nations Conference on the Human Environment in Stockholm in 1972, the establishment of the United Nations Environmental Program (UNEP) was decided, and soon the building of its headquarters in Nairobi started. In the following years, desertification in Africa was targeted by UNEP as its most important focus area. Subsequently, the UN's Desertification Conference was organised in Nairobi in 1977. Two years later one of UNEP's directors followed up by proclaiming that desertification represented the greatest single environmental threat to the future well-being of the Earth.

Swift (1996) has explained this choice of focus by the fact that the desertification narrative was viewed as non-political and that it did not have any powerful losers. The villains in this Malthusian story were small-scale farmers and pastoralists who did not have a voice in global environmental political debates.

Twenty years after the Stockholm conference, the Rio UN Conference on Environment and Development again highlighted desertification as a global problem, and it was decided to establish a UN Convention to Combat Desertification, which was subsequently adopted in 1994. Repeatedly, the Sahel has been presented as the world region most at risk to be affected by this process.

After the droughts in the Sahel in the 1970s and 80s, there was an international focus on overpopulation as a main cause of desertification. This Malthusian understanding of the issue dominated until the early 2000s when climate change was mounting on the global political agenda, which not only led to a securitisation of this problem, but also sparked a new interest in the issue of desertification. This time the idea emerged that the process was mainly driven by climate change.

Scientific research has, on the other hand, given a completely different picture of changes in vegetation and landscapes in the Sahel. Soon after the big drought in the mid-1980s, research in remote sensing using data from satellites indicated a fast recovery of the vegetation (Hiernaux 1988, Tucker et al. 1991). These trends were later confirmed by more studies (Eklundh and Olsson 2003, Olsson et al. 2005, Anyamba and Tucker 2005, Herrmann et al. 2005, Fensholt et al. 2006).

The re-greening of the Sahelian landscape was detected both in the grass cover as well as the woody vegetation. For instance, during 2000–2014, woody cover increased substantially in the Sahel (Brandt et al. 2016). Most of this greening can be explained by increased rainfall. However, farmer-induced improvements on farmland through tree planting may have locally contributed to this greening (Reij et al. 2005). In other words, the overall positive trends observed for both grass and woody plant cover question the desertification narrative in the Sahel. Moreover, this rapid recovery of the vegetation after droughts has also been observed in other drylands around the world, as noted by the IPCC (Mirzabaev et al. 2022).

COLONISATION, DESICCATION AND DESERTIFICATION

The idea that the Sahara Desert is spreading south is, however, not new. Already in the 1880s, this idea emerged among colonial administrators and scientists in tandem with the colonisation process in the Sahel. From the beginning of European presence in the region, there was a firm colonial belief that the local African population was creating desert-like conditions through overuse of natural resources such as pastures and forests. At the same time, there was a view among some French scientists that the region was subject to climatic desiccation (Benjaminsen and Hiernaux 2019).

This desiccation was seen as a process of diminishing rainfall and a generally drier climate. The early colonial period in the Sahel (late 19th and early 20th century) coincided with the beginning of a dry period. As a result of drought, in addition to grasshoppers, rinderpest, and not the least, military occupation, there was economic decline and hunger in the Sahel at the onset of the 20th century. These years of drought increased a concern among French administrators and scientists that the Sahara was spreading south, which again represented an obstacle to reaching production targets in the colonies.

In France, this concern led to a debate about whether desiccation was a natural or human-induced process. At the Colonial Exposition in Marseille in 1906, the geographer M.J. Lahache launched the thesis that a gradual and natural desiccation of West Africa was taking place (Lahache 1907). In his view, drought and potentially desiccation represented a serious obstacle to the prosperity of the French colonial empire in Africa. The cause of this desiccation was either a change in the radiation from the sun or in the wind patterns, which again resulted in decreasing rainfall. But Lahache also took a cautious stand; he added that there was not enough data as yet to ascertain that desiccation was actually occurring.

The geologist and colonial administrator Henry Hubert became one of the most vocal French defenders of a desiccation hypothesis (Hubert 1917, 1920). He was less cautious than Lahache and firmly argued that the desert was advancing at a rate, which called for swift action. Hubert believed that desiccation was a global process occurring also in other world regions. In West Africa, he concluded that a progressive desiccation had been going on for 60 years with an aggravation in the last 20 years. Hubert saw both natural and human factors behind the on-going desiccation process, but with decreasing rainfall as the main factor. To combat this desiccation process, he recommended a general fight against erosion and the invasion of sand through protection of the vegetation and systematic reforestation.

The geologist René Chudeau was, however, critical to the idea of gradual desiccation of the Sahel, which led to a standoff between Chudeau and Hubert (Van Beusekom 1999). Chudeau argued for fluctuations as the dominant trend instead of desert advance. He did not relate this trend to local resource use, but studied rainfall patterns in West Africa to answer the question of whether there was a progressive desiccation taking place as claimed by Hubert.

Chudeau presented rainfall data from Saint Louis (Senegal) from 1830 to 1915 to support his fluctuations thesis. In the same vein, he referred to variable water levels in Lake Chad, concluding that West Africa in 1921 was in a dry period with a slight improvement during the last few years, and that there were no proofs of desiccation. The data would rather support the idea of 20–50 year cycles, he stated (Chudeau 1921).

While some scientists had, in the first couple of decades of the 20th century, argued that desiccation was a natural process, this view was increasingly replaced by the idea that humans and their livestock were to blame, which from the 1920s was further strengthened and became part of the colonial ideological position mutually supporting desiccation theory (Davis 2016a). Many administrators and scientists strongly believed that native Africans were generally mismanaging natural resources leading to deforestation, soil degradation and desert advance. In particular, these opinions prevailed among colonial foresters.

For example, in 1907 an expert group commissioned by the French Ministry of Agriculture to assess the state of forests in French West Africa had expressed the view that desiccation was mainly caused by local human activities. One aim of their mission was to propose appropriate action 'considering the importance to take measures to preserve the forest masses of the African interior and to assure the reforestation of the naked regions'.[1]

J. Vuillet, an agronomist and head of the Agricultural Service in Upper Senegal and Niger, headed the Forestry Mission. Vuillet was less convinced about Lahache's theory on natural desiccation. Instead, the focus of the report was on various forms of local overuse and destruction of forests. The main assumption and starting point of the Mission was that extensive deforestation was taking place in West Africa. By personal observation, the members ascertained that 'the forests are in an appalling state' and that 'the causes of deforestation in the territories covered by the mission were: 1. Clearing of new land for cultivation; 2. Bush fires; 3. Grazing and browsing; 4. Cutting of trees for wood use'.

Referring to bush fires and overgrazing, the report continued: 'In this way, the ruin of the forest is consumed under the eyes of careless and almost irresponsible populations.' 'It is indeed true,' the Forestry Mission concluded, 'the Sahara progresses toward the South; and that because of Man's action.' The use of bush fires to manage pastures is called 'a barbarian practice' and deforestation and desert advance is caused by 'the indolence and carelessness of the blacks'.

The forest service, which was proposed by Vuillet and his team, should introduce scientific methods in forest management in the colony, because 'any forest exploited according to the rules of forest science does not die, (but) continues to exist, is improved and gives every year a number of products,' while 'any forest, which is exploited without any method and abandoned to fires, to beasts and to the devastation by men, is fatally condemned to disappear in the short term'. The proposed forest service would be authorised to conserve, improve, police and exploit forests and to restore the parts of the forests which were in ruin.

The next major Forestry Mission to French West Africa was organised in 1923 (Mangin 1924). The head of the Mission was M. Mangin, inspector of the Forest Service in France. The aim was to identify the state of the colonial forests and to propose an organisation for the conservation, development and management of the forests. Following this double objective, the report is divided in two with the first part giving botanical descriptions of various forest stands in different parts of West Africa. The second part discusses 'la question forestière':

> The native ... who is used to granting no price on what he obtains without labour has always considered the forest and still considers it as a good without value and besides inexhaustible. He has misused it and still abuses it without perceiving the consequences of his lack of foresight.

While the Vuillet Mission had proposed the establishment of a forest service in French West Africa, nothing had as yet been done. This may have been the result of a certain resistance to forest conservation among colonial officials (Fairhead and Leach 2000). Administrators relied on local chiefs for tax collection and recruitment of corvée labour, and introducing widespread restrictions on forest use, cultivation, grazing and bush fires could compromise this cooperation that the colonial administration depended on. In addition, forest issues continued to be dealt with by the same agents handling agriculture and livestock production within the Agricultural Service, and these agents also feared that strengthened forest conservation through establishing a forest service could contradict agricultural development. Colonial development was also a major consumer of fuelwood for steam engines and wood for construction, which may also have contributed to the hesitation among colonial officials in establishing a forest service.

In order to fight against the alleged severe deforestation in West Africa and to circumvent local colonial politics, the Mangin Mission argued for the establishment of an independent forest service that would report to its own head in Dakar instead of to the local Governor. The Mission also proposed making a new law to replace the Forest Decree of 1900.[2] The legal inaccuracy and the insufficiently defined implementing tools of this decree made it incapable of dealing with the severe forest depletion taking place, the Mangin Mission report argued.

The follow-up of the Mangin Mission finally led to the establishment of the Sahelian forest legislation most frequently referred to as the forest decree of 1935.[3] This law also provided for the creation of the Water and Forest Service (Le Service des Eaux et Forêts) – often referred to only as 'the forest service', inspired by a similar organisation in France, and mandated to implement the forest policy, which recruited its agents from the military and the police. This role of foresters as policemen rather than extension agents continued through the colonial period and persisted after independence (Brinkerhoff and Gage 1993, Benjaminsen 1997, Becker 2001).

From the 1930s, desiccation theory was gradually replaced by a more explicit focus on desertification. This transition was in particular inspired by the dust bowl event that hit the southern Great Plains in the United States in the early 1930s (Van Beusekom 1999). Aubréville (1949) is incorrectly said to have coined the term 'desertification', for already in 1927, the French forester Louis Lavauden, who first worked in southern Tunisia and later became head of the colonial Forest Service in Madagascar, had introduced this notion (Lavauden 1927). Lavauden had also participated in one of the first trans-Saharan motorised expeditions in 1925 from Tunisia to Lake Chad and further to Dahomey and in this way observed both sides of the Sahara Desert. He stated that 'desertification' was a relatively new phenomenon, which was exclusively a result of human action.

To arrest this desertification process north of the Sahara, a strict forest and pastoral policy was necessary, he said, including reforestation, restrictions on grazing, fighting against bush fires, and establishing areas reserved for preservation without use. However, south of the Sahara, this question was even more difficult, because one had to deal with 'primitive populations, incapable of understanding the utility of rules'.

In the 1930s, the British forester E.P. Stebbing was one of the most well-known and active scientists writing about 'the threat of the Sahara' (Swift 1996, Grove 1997, van Beusekom 1999). 'His perceptions of the dry season Sahelian landscape provoked him into writing a feverish warning on what he saw as the dangers of desertification' and he widely publicised his thoughts through a number of lectures and publications (Grove 1997). In his first report, which was given as a lecture to the Royal Geographical Society in London, Stebbing stated that '... in this region the population is actually increasing whilst the means of supporting it are obviously and visibly decreasing' (Stebbing 1937). To stop the progress of the Sahara he proposed to reserve two parallel forest belts through the French and British colonies, which should be 15 miles deep and 1370 miles long. These two belts would be closed and protected from farming, fire and grazing.

To further substantiate the facts of desert advance, Stebbing proposed the establishment of a commission to investigate the matter further. The following year, the Anglo-French Boundary Forest Commission was appointed, and its members travelled in the border area between Niger and Nigeria in 1936–37 to investigate the question of desert advance. Taking Stebbing's description as a starting point, the mandate of the Commission was to interview people

on both sides of the border and to gather information regarding the desiccation process, the extent to which the forest had lost ground to the steppe, the
extent of the new desert regions, and the causes of this new situation. However,
the Commission carried out its mission just after an exceptionally wet rainy
season. The team members were told that people had not seen that much rain
in living memory, which gave them a completely different image than the one
presented by Stebbing.

In its report, the Commission concluded that it had not been able to observe
any massive displacement of sand. 'Hence, the cultivated land in Northern
Nigeria and Niger is at present not threatened by any invasion of sand' (Anglo-
French Commission 1973). Natural regeneration was observed to be abundant
and, in many places, assisted by livestock spreading the seeds. In some places,
the inhabitants even confirmed the team members' impression that the areas
visited had been less forested earlier. The mission was also not able to support
the idea of a movement of people from the north to the south. On the contrary,
it observed that sedentary farmers were spreading north. The conclusion of
the report was that there was no immediate danger of desiccation and desert
advance in the areas visited.

In parallel to this debate about causes of desiccation and desert advance,
there had also been critical voices among French scientists who were present-
ing more optimistic viewpoints on the environment in the Sahel, in addition to
René Chudeau who has already been mentioned.

Another critical scholar was the agronomist Georges de Gironcourt. In
1908–09, he went on a scientific expedition to the Gourma area south of the
bend of the Niger River in French Sudan. Based on his personal observations,
he argued against the idea of desert advance and that local resource use was
degrading the land (De Gironcourt 1910, 1912). Instead, he argued that there
were no data to support such an idea. On the contrary, the vegetation and the
rains seemed to fluctuate and to recover after a drought anticipating hereby
one of the main ideas in present day critiques of the desertification narrative.

In a public lecture given in 1913, he rather blamed deforestation occurring
in some places on the colonial government:

> Deforestation has been more active during our ten years of occupation than during
> several centuries of indigenous nomadism. The use of wooden frameworks for the
> construction of our stations leads to the felling of innumerable palms and the heat-
> ing of our steam engines consumes an amount of wood, which is not in proportion
> to the woody production of the banks of the Niger.[4]

Contradicting frequently heard views on the scarcity of resources in the Sahel,
he also argued for a more optimistic vision of the development potential along
the Niger river and that the inner Gourma contained pastures, which could

support considerably larger herds of livestock than what was the case at the time.

The colonial narrative that de Gironcourt had criticised in 1910 had already started a gradual transition from a focus on desiccation to one on human-induced desert advance. Aubréville's paper from 1938 on 'the colonial forest' demonstrates this transition where he first asks in the introduction whether desiccation is caused by natural causes or human-induced deforestation. Answering this question, he concluded:

> Deforestation results from the systematic destructions operated for centuries by indigenous people. The climate has changed little since antiquity. The forest flora, narrowly controlled by the climate, has remained the same, at least where humans had not exerted their influence (Aubréville 1938: 42).

Aubréville, who was the first head of the Forest Service in the French colonies, further strengthened the crisis narrative by blaming degradation on local land-use practices. This helped justify the Forest Service taking control over forest resources and imposing rules on their use. In this institutionalisation and the enhancement of the idea of human-induced desertification, the forest climax model became useful.

This climax model, first developed by the American botanist Clements (1916), provided theoretical support to the idea of a stable 'climax state' of the vegetation in equilibrium within a pristine environment and climate. The model views the vegetation dynamics following a disturbance as a unidirectional sequence of stages, named succession, ideally leading back to the climax, unless disturbances were too strong or repetitive, such as with 'local mismanagement' impeding the return to equilibrium and ending in a desert-like landscape.

In this crisis narrative, local people's values and natural resources management practices were increasingly depicted by colonial policies as backwards, unproductive, and deleterious to ecosystems (Fairhead and Leach, 2000). Shifting cultivation, bush fires for different purposes, and pastoral management in particular were criminalised through colonial policies and legislation.

While there was drought in the Sahel in 1913–14 and during 1930–32 and 1944–48, the 1950s were exceptionally wet with abundant pastures and regrowth of forests. This also meant little focus on desertification in the first decades after the Second World War. However, when drought hit again in 1972–73 and 1983–84 the desertification narrative was quickly reactivated. Scientists inferred that drought demonstrated the vulnerability of Sahelian ecosystems resulting from increasing population pressure (Boudet 1972, Peyre de Fabrègues 1984, Sinclair and Fryxell 1985).

The media attention following these great droughts in the Sahel in the 1970s and 1980s brought through TV broadcasting the combination of environmental degradation and hunger into the living rooms of people in the West. Many Western NGOs became involved in emergency aid to help those in need. After several years of aid, many of these projects became long-term development projects that generally shared the concern for the Sahel's environment and the view that the land-use of the local populations was responsible for the crisis.

This increased global attention to desertification coincided with an international environmental movement in the 1980s and the rise of the concept of 'sustainable development' (WCED 1987). In 1982, Mali agreed on the implementation of a Structural Adjustment Programme with the World Bank and the International Monetary Fund. This led to a reduction in government spending and in the number of civil servants in all sectors except in forest management where the number of forest officials in rural areas throughout the country increased many-fold from the mid-1980s (Poulton and Ag Youssouf 1998). This led to an increased presence of forest agents to police village and pastoral land, which was seen by the government and international aid donors as necessary to foster sustainable development through stopping an advancing desert in effective ways.

DEPICTING DESERTIFICATION

In disseminating the desertification narrative, images have played a key role. In environmental discourse analysis, various discursive elements have been analysed since the emergence of this field in the early 1990s with narratives and metaphors as rhetorical devices occupying a central role (Hajer 1995, Dryzek 1997, Adger et al. 2001).

In addition, images (drawings, landscape photographs, maps) are often used as forms of non-verbal rhetorical devices. Büscher (2016) refers to the increased use of social media by environmental organisations, which has rapidly changed the political economy of conservation and led to a reimagining of nature and human-nature relations. These new relations include human alienation from nature combined with remote environmental engagement through the use of new media, such as conservation websites, blogs, videos, photos, text, and online chat forums. For instance, Igoe (2017) studied the use of romanticised and spectacular online photographs and videos of African wildlife and wilderness to sell safari destinations as well as conservation interventions, but which adversely result in alienation and commodity fetishism (see chapter 5 for more on commodity fetishism).

However, this chapter does not focus on what is romanticised and subsequently appropriated and commoditised for external consumption, but on what is demonised in order to be appropriated by competing land-uses. Nonetheless, these two contrasting tropes may also be combined, as in the case of the romanticised African wilderness in tandem with claims that pastoralists over-stock and represent a threat to this wilderness.

In discourse analysis carried out within environment and development studies as well as in political ecology, Emery Roe's approach to narrative analysis with a focus on crisis narratives has been influential (Roe 1991, 1999). Roe is especially concerned with arguments and courses of action in a narrative. In addition to such storylines, narratives also tend to contain a gallery of actors, such as the archetypical heroes, villains and victims (Svarstad 2009).

Narratives are stories with a beginning, middle and an end, or when cast in the form of an argument, with premises and conclusions. According to Roe (1999: 6):

> crisis narratives are the primary means whereby development experts, and the institutions for which they work, claim rights to stewardship over land and resources they do not own. By generating and appealing to crisis narratives, technical experts and managers assert rights as 'stakeholders' in the land and resources they say are under crisis.

In a development context, this means that crisis narratives have portrayed marginalised people as the villains (and usually also the victims of their own mismanagement), while external experts associated with state policies are pre-sented as the heroes who provide the solutions. In this way, states and other powerful actors have used such narratives to assert authority over land and natural resources at the expense of customary land-users (Leach and Mearns 1996, Stott and Sullivan 2000, Bassett and Zuéli 2000, Adger et al. 2001).

To help us conceptualise the impact of images within environmental narratives, the classic work of Roland Barthes may be useful. Barthes (1961 [1977]) articulates visual communication into the two separate levels of denotation (literal meaning of an image) and connotation (societal values communicated through an image). He added that image and text are closely connected, and that image does not necessarily illustrate text, but that it may often be the other way around. The caption, in particular, is included in the image denotation.

Later, Barthes (1972) added myth as a third layer, which refers to historically grounded ideological ideas that are evoked by an image. Myth is depoliticised speech that 'has the task of giving a historical intention a natural justification, and making contingency appear eternal' (Barthes, 1972: 254).

However, it will often be difficult to separate in practice these three levels of meaning, because interpretation will vary with cultural background and also because the levels may overlap to some degree (Chandler 2017). Nevertheless, the often repeated images of cracked earth in Figure 2.1 may tentatively serve as an example. The denotative meaning of this type of image would simply be

Note: This is usually soil which cracks after having been flooded. Clay is often found in the lower parts of the landscape in the Sahel. These areas are flooded in the rainy season before they dry up in the dry season. When searching Google Images for 'image desertification', various similar photos of cracked clay are offered. Especially around the annual World Day to Combat Desertification (17 June), such images accompany articles on desertification published on websites and in printed media on a global scale.
Source: Wikimedia Commons.

Figure 2.1   Cracked clay as an image of desertification

a dry landscape with cracked soil, and the connotative meaning could refer to nature under threat through overexploitation. The mythical meaning would be related to the historically manufactured idea of 'desertification'.

The cracked earth images in Figure 2.1 represent an iconic image that efficiently transmits ideas of desertification conveying variegated associations to its viewers depending on their knowledge and background. Yet, among most observers, these photographs will generate the idea of desert encroachment (in Africa) caused by local mismanagement or climate change, or both in combination.

Myths move beyond connotation, because they are taken for granted and not questioned. In this way, myths have much in common with crisis narratives that also resist empirical knowledge that goes against their storyline (Roe 1999). As Barthes points out, myths are depoliticised stories that work to conceal the ideological function of images. In Figure 2.1, the hidden ideology would be a Eurocentric, Malthusian and neo-colonial view of African landscapes and their users, thereby contributing to the justification of the use of strict measures to arrest desertification – through fines, restricted use, and sometimes the use of violence.

Various images may also be used as powerful tools to 'problematise' (sensu Li 2007) indigenous pastoralism, whereas technical measures, such as implementing carrying capacities or planting trees, are presented as solutions eventually leading to pastoralists being dispossessed of grazing land.

Aubréville (1949) represents an early expression of the method of depicting historical landscape change in the Sahel (Figure 2.2). He simply drew the changes he claimed were occurring.

Figure 2.2 is an example of a drawing that was included in his seminal book 'Climate, Forests and Desertification in Tropical Africa' (Aubréville 1949). These drawings were intended as illustrations of on-going changes in forest and savanna ecosystems at the time. Aubréville clearly saw the drawings as important to transmit the message of ecosystems under threat from local land-use practices. They are of a large size in a format that can be folded into the book. At three places in the lower drawing, Aubréville's perceived indications of the degrading effect of bush fires (feux de brousse) can be seen. Nevertheless, at this point in time, desertification had not yet attained the status of myth, although Aubréville was one of the early influential academic contributors to establish desertification as an iconic environmental threat.

When the image of drying or of human-induced desertification is projected in international media or by development organisations, iconic photos of cracked earth (Figure 2.1) are typically used, or alternatively of a sand dune threatening a single tree or a cultivated garden. Presentations of such image-tropes are rarely specific about location, since time series from specific areas will tend to support the greening counter-narrative. While such images work to offer visual examples of ongoing degradation, they paradoxically remain abstract representations that cannot be located in space and time. Hence, the geographical and historical context of these images is rarely provided to the viewers.

Images of desertification currently serve to justify a continued fight against this perceived process through fundraising for large-scale tree planting projects, such as the Great Green Wall in Africa (chapter 5). Images of green trees

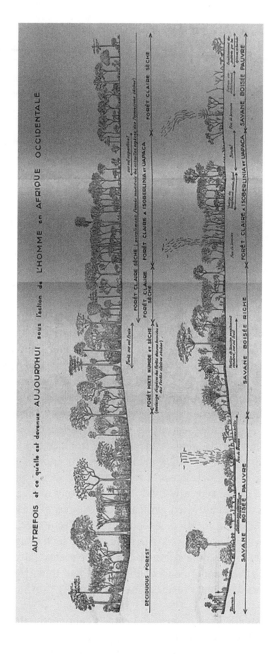

Figure 2.2   *Illustration of forest landscapes in West Africa according to Aubréville (1949)*

planted in the desert clearly still hold sufficient visual power to resonate widely in the global media. The connotative level of these images is to communicate that science and modern principles of sustainability, usually introduced by foreign or external experts, can save the environment from overexploitation and disaster.

For instance, the BBC can report that 'Many of the social and economic problems in the Sahel can be linked to a process called desertification' and that:

> Overgrazing and deforestation are the primary causes of desertification and have turned vast swathes of the Sahel into dust. However, an ambitious afforestation programme known as the Great Green Wall aims to reverse this process (The BBC 3 May 2013).

Therefore, planting trees becomes a universal remedy to problem-solving in the Sahel. The BBC even speculates whether such large-scale tree planting can help increase the security in the region and undermine recruitment for 'terrorist' or 'jihadist' armed groups. This belief in desertification being one of the causes of the current insecurity in the Sahel is frequently repeated in popular media and policy documents (including in a documentary film produced by the Great Green Wall project).

Hence, establishing green walls also becomes a way to depoliticise and to 'render technical' (Li 2007) a deeply political problem about governance, as well as access to land and natural resources. Consequently, mythical images of desertification blaming pastoral land-use help to justify the dispossession of pastoralists whether through large-scale agricultural schemes or afforestation programmes.

FOREST GOVERNANCE

The Forest Service has been at the centre of rural people's negative perceptions of state services and corruption for several decades. During the increased international attention on 'desertification' as a global and, in particular Sahelian, environmental problem after the droughts of the 1970s and 80s, the Forest Service in Mali was given a wide mandate in addition to new funding (Benjaminsen 1993, 2000, Ribot 1995).

Such a mandate was believed to be necessary because of a widely shared narrative of a fuelwood crisis in the Sahel that needed urgent attention (e.g. Eckholm et al. 1984). This narrative also had a strong position in Mali in the 1980s and 90s expressed, for instance in the National Plan to Fight Desertification (Direction Nationale des Eaux et Forêts 1985). The narrative focused on a growing gap between diminishing resources and an increasing demand for fuelwood. The French-funded TransEnerg project, for instance, predicted that all stocks of wood around Bamako would be finished by 1990 (Bertrand et al. 1984).

The narrative was, however, questioned by empirical studies in the Sahel and other African drylands (Leach and Mearns 1988, Cline-Cole et al. 1990, Benjaminsen 1993, 1997). The impact of fuelwood harvesting on forest areas is difficult to estimate. First, an important part of the wood that is gathered originates from land that is cleared for farming. Second, another significant part of the wood consumed comes from dead trees. This is especially the case in the northern Sahel where the drought led to increased mortality of the woody vegetation, which improved people's access to fuelwood in the short and medium term (Benjaminsen 1996). Further south, even in the more densely populated and intensively used areas such as the cotton zone in southern Mali, about half of the household energy consumption came from dead wood in the mid-1990s (Benjaminsen 1997). Urban areas are, however, supplied mainly with fuelwood coming from green wood, although Gautier et al. (2013) concluded that no shortage of supply had been observed as yet.

The consequences of these influential ideas on desertification and an associated fuelwood crisis were a top-down approach based on handing out fines to people for activities that were seen as leading to environmental degradation and desertification. This often meant fining women for collecting dry wood and households for not having wood-saving stoves. Pastoralists, who generally do not have a problem finding dead wood for cooking were, from the mid-1980s in particular, harassed by armed forest agents who would visit nomadic camps and give people fines for one reason or another (e.g. for collection of firewood or wood for handicraft, or for not having energy-saving stoves) (Benjaminsen 1993).

Through these practices, forest agents became a constant threat to rural people and their livelihoods all over the country leading people not to see the Forest Service as defenders of nature, but as money-grabbers (Sanogo 1990). Forest agents were allowed to keep a portion of the fines collected, in addition to all the fines they collected informally without receipts, which further encouraged their rent-seeking behaviour leading the Forest Service to become a vehicle for 'decentralised plunder across the country' (Poulton and Ag Youssouf 1998: 27) and resulting in foresters topping rural people's hate list.

In Mali, the Forest Act of 1935 was revised first in 1968 with only a few small changes and then again in 1986. The previous president, Moussa Traoré, who seized power during a coup d'état in 1968 and lost it during riots and a new coup d'état 23 years later in 1991, became, during the emergence of the environmental movement in the 1980s, a convinced 'environmentalist'. To impress aid donors and environmental organisations and to attract more environmental aid to the country, the Forest Act was made even stricter in 1986

(Ribot 1995). Extremely high fines compared to the level of income in Mali were introduced.

The Forest Act of 1986 also prohibited all use of bush fires. If foresters discovered burnt vegetation on village land, the village risked receiving a common fine of up to US$1,000. Some villages depended on burning to the extent that they would take the fine rather than stop the burning.

In the southern Sahel, fires are used as part of traditional management of pastures (Laris 2002). Farmers usually start a yearly cycle of bush fires early in the dry season to fragment the landscape in order to prevent later fires that can damage the local environment. This creates a seasonal mosaic of habitat patches that increases the sustainability of a variety of landscape uses (Laris 2002).

Laris (2002: 155) also argues that the dominant discourse on African savanna burning 'overemphasises the ecologically detrimental aspects of fire, while neglecting the beneficial ones resulting in misguided policies that pose a threat to human livelihoods and savanna ecosystems'.

Hence, rural people consider burning of grazing land at the various parts of the year to be advantageous for bushes and perennial grass. In the northern part of the Sahel, however, people do not use fire in the management of pastures because the grass cover there mainly consists of annual species (Benjaminsen 2000).

This traditional knowledge on the use of fire was, however, neglected by the Forest Act. President Traoré even made the battle against fires one of his pet subjects. Ironically, the Koulouba Hill, where the president's palace is located, was burnt every year at the end of the rainy season. When it was lighting up the night sky and in good view of the whole capital, it represented a clear protest against Traoré's environmental policies (Benjaminsen 2000).

The Forest Act of 1986 required that all households used so-called 'improved stoves' to save on wood. This order was based on the idea that the local use of wood for food preparation, representing about 90 per cent of all use of energy in Mali, was a threat to the forests. Starting in March of 1987, people who did not have such a stove risked being fined about US$15. For large parts of the rural population, such an investment of a stove was unnecessary, because they had more than enough dead wood and branches to use (Benjaminsen 2000). In addition, it was not possible to move the most common and cheapest wood-saving stoves that were built of clay. This type of stove is also not very durable and collapses during each rainy season. It also gives off less light and heat in the cool northern Sahelian winter nights, because the stove, for it to be effective, has to fit tightly around the pot.

Those who gained from these strict laws were the Forest Service who had fines as their most important source of income (Benjaminsen 1993). They were formally allowed to keep a certain percentage of the money collected.

In addition, there were fines that were never officially reported. In that way, it became lucrative to be a government-employed forester. As we shall see in chapter 3, the heavy-handed forest governance that was strengthened through international support in the era of sustainable development in the 1980s forms the backbone of the general condemnation of rent-seeking government officials, which again helps recruit rural people to jihadist groups.

NOTES

1. All translations from French to English have been done by the author. This quote is from the report of the Vuillet Mission available at *Archives nationales d'outre-mer* (Aix-en-Provence), Affaires Agricoles R24 (14 MI 1566).
2. The very first law relating to the use of forests and forest products in the French part of the Sahel was the decree of 20 July 1900 for Senegal and Dependencies. It allocated restricted use rights to native populations to collect nuts and wood, to hunt and to use the forest areas as pastures. However, if the use of the resources jeopardised 'the richness of the forest' the Governor would take the necessary protective and prohibitive action (article 23). Apart from these restricted use rights, the decree introduced permits for commercial extraction of forest products and fines for infractions of the rules.
3. *Journal Officiel de l'Afrique Occidentale Francaise*, 3 August 1935, pp. 611–8.
4. Reported in *Journal Officiel de la République Francaise*, 3 February 1913.

3. Pastoralism, moral economies of resistance and jihadism

PASTORALISM AS A LIVELIHOOD STRATEGY

Pastoralism is a major livelihood strategy all over the Sahel. This is an extensive livestock-keeping system based on the herding of animals often over long distances and in some of the most remote and marginal environments on the planet. Globally, mobile pastoralists number about 200 million households and use about 25 per cent of the total land area (Dong 2016).

Pastoralists often operate in non-equilibrial environments that are unstable, fluctuating and uncertain. These environments are driven more by climatic variation than livestock numbers and grazing pressure (Behnke et al. 1993) – typical examples are grazing areas in the dry tropics (Sandford 1983, Turner 1993, Sullivan and Rohde 2002, Benjaminsen et al. 2006, Hiernaux et al. 2016) and pastures in the Arctic (Behnke 2000, Benjaminsen et al. 2015, Marin et al. 2020).

Pastoral strategies generally revolve around the principles of flexibility and adaptation to fluctuating climatic conditions through the various seasons, and not the least through environmental and climatic variations from one year to the next. Through their extensive land-use, pastoralists generally do not claim, or need, exclusive property rights to land. They rather depend on access or use-rights at certain times of the year. In many pastoral areas, especially along migratory routes, there is overlapping use in co-operation with other users. In some areas, as in parts of the Sahel, co-operation has been forced through the threat of violence as pastoralists have been dominating farmers through military means. When co-operation fails, or when military domination becomes in flux, conflicts emerge (chapter 4).

Pastoralists have through generations obtained practical experience and knowledge to cope with uncertainty and variability mainly through a mobile and flexible approach (Krätli and Schareika 2010). Through this mobility and flexibility, pastoralists may be better able to adapt to a changing climate than other land-users (Davies and Nori 2008, Krätli and Schareika 2010, Jones and Gutzler 2016).

They also have a considerably lower climate impact than other livestock-keeping systems, in particular stall-feeding and ranching. For instance, pastoralism in the Sahel may be seen to be carbon neutral (Assouma et al. 2019). Policies of settling pastoralists in villages and limiting migrations may therefore lead to land degradation and increased emissions of climate gases.

Modern states have indeed tried to settle pastoralists and confine their movements within clearly defined boundaries under the argument that their land-use is neither ecologically sustainable nor economically productive. States also have a tendency to view pastoral use as messy and illegible (Scott 1998), since migrations do not fit into a modern form of governmental organisation. Pastoralists are difficult to tax and control, and it is also difficult to adapt services such as schools and health care to their mobility. The consequence has been that pastoral cultures, way of life and land-use have been under pressure from an expanding, sedentary form of modernity for many decades.

Furthermore, pastoralists are usually minorities in the countries they live and have often been particularly prone to marginalisation and disenfranchisement as a result. Since they occupy marginal lands that, for historical reasons, tend to be areas where there are international borders, they sometimes represent, or are represented as, a threat to the integrity of the modern state.

In addition comes new pressures on rural land in general and pastoral land in particular, which are the result of the confluence of several global trends. (1) Rising food prices in 2007–08 led important international investors to 'rediscover' the agricultural sector. This has also led to growing pressure on pastures in drylands and the conversion of these pastures to farmland, especially in cases where water for irrigation is available. (2) Increasing energy prices have led to the conversion of pastures for biofuel production. (3) Projects to mitigate climate change in the form of afforestation have also led to the exclusion of herders from pastoral land. (4) Rising prices for minerals, such as copper and gold, as well as fossil fuels, have led to plans to open new areas for exploitation. (5) Increasing initiatives to conserve biodiversity in remote 'wildernesses', which have historically made significant contributions to pastoralists' displacement and experiences of injustice, (e.g. Brockington, 2002) continue to result in the exclusion of pastoralists in specific, contemporary cases, often through the introduction of 'ecotourism' (Benjaminsen and Bryceson 2012).

In the Sahel, it is primarily the rain that decides whether the grazing will be good or bad during a year. But even in good years, pastoralists depend on moving to where water and grazing are available.

In addition to moving, most pastoralists in the Sahel choose to spread risk by combining their herds consisting of different types of animals (goats, sheep, cows, camels). It is risky to put all of one's eggs in one basket in the drylands. Each type of animal has its own needs for grazing and water. People who depend on natural resources in such arid regions have therefore learned that the only certainty is uncertainty (Berge 1999).

Sahelian pastoralists have, however, for decades complained about a continued loss of access to dry season pastures and to livestock corridors. This process has been facilitated by a general negative policy discourse surrounding Sahelian pastoralism among governments and aid organisations (Thébaud and Batterbury 2001, Pedersen and Benjaminsen 2008).

Also, the French colonial government felt that it was important to settle pastoralists, not only for purposes of administrative or ecological control, but more generally to promote development.

This view was clearly expressed in the Forest Act of 1935, which distinguished between 'land where value has been added' and 'empty' or 'unused' land. Cultivated land was included in the first category, and farmers were given use rights to cultivated land and fallow land that had been used less than five years. Southern Mali with its more fertile and productive farmland was described as 'the useful districts' by the French. Grazing land and forests, on the other hand, were defined as unused land with no recognised use rights.

For administrative and ecological reasons, the colonial government drew lines within districts, so that each pastoral group and each village was given their own territory where they were required to stay. The French drew up territorial borders in a way that pastoralists were excluded from access to key resources. The objective was to weaken the pastoralists militarily and to reduce the risk of armed resistance. The borders were also meant to hinder different groups of pastoralists from coming too close to each other or to villages, and were therefore meant to prevent conflict and pillage. It was illegal for pastoralists to cross the borders without permission from the colonial administration (Berge 1999).

In their own justification, the French colonised to civilise. In the colonial archives, there is an expression that is repeated again and again: 'La mission civilisatrice de la France' (the civilising mission of France). Reducing friction and conflict between ethnic groups through administrative control and division was seen as a central part of this mission.

In addition to moving and spreading risk, the swings in the accessibility of pastures and food for pastoralists are curbed through storage. While farmers base their adaptation on storing grain from one harvest to the next, Sahelian pastoralists have 'stored' by having larger herds of livestock than they immediately need, in order to sell animals on markets or trade animals for grain with farmers during the dry parts of the year.

Outside experts have often argued that pastoralists should sell when the animals are fat and the prices are high, rather than wait for the dry season. Pastoralists, however, often wait to sell animals until they have a special need for cash, for example, because of illness in the family. If they do sell animals, they often prefer to keep the surplus in the form of items that are light compared to their value, such as jewellery.

Moreover, the contribution to the national economy of pastoralism in the Sahel is often not understood. In Mali, the export of livestock to neighbouring coastal countries was for many years the second-largest export commodity after cotton. In recent years, livestock has dropped to third place after gold and cotton.

In fact, the drylands of the Sahel have a comparative advantage when it comes to livestock production, but this advantage depends on the continued possibility to migrate between the rainfed wet season grassland on the savanna and the dry season pastures in wetlands. Research on human nutrition in the Timbuktu region has also shown that children of the specialised nomads are better fed than those of farmers in northern Mali and are at the level of children from the more productive agricultural areas in the southern parts of the country (Hatløy 1999).

Finally, it is important to note that labour still tends to be a minimum factor in agrarian systems in the Sahel, despite high human fertility rates. It is therefore often difficult for the specialised pastoralists to combine herding with farming, because in order to be able to herd livestock as a way of living, it is necessary to be able to move quickly to where pastures are available. Combining pastoral migrations and farming requires more manpower than most households in the northern Sahel can procure (Pedersen and Benjaminsen 2008). Such labour scarcity was solved in the past by pastoralists taking slaves in the sedentary farming population.

Despite a certain scarcity of labour, combining farming and herding within agropastoral adaptations is common further south in the Sahel. But the migrations involved are usually short except in large or wealthy households that can hire herders.

MORAL ECONOMY

Pastoralists have, since 2012, continued to join the various jihadist groups in the Sahel and tend to be in the majority in these groups (Benjaminsen and Ba 2019). They are in general attracted by a pro-pastoral, anti-state and anti-elite discourse among jihadist leaders. Over several decades, the rural peasantry, and especially pastoralists, have become increasingly fatigued by and disgruntled with a predatory and corrupt state, which extracts rent from the rural peasantry. In addition, the development model imposed by the state and the aid industry has not responded to pastoral priorities and instead gradually dispossessed pastoralists of access to livestock corridors and dry season pastures.

Examples of rent-seeking include land-use conflicts not being resolved because local officials and judges have received payments from both parties to support their claims, leading to a lack of adjudication (Benjaminsen and Ba 2009), as well as the practices of the Forest Service already mentioned in chapter 2, for instance, randomly fining women collecting fuelwood and herders grazing livestock. The result of this growing anti-government sentiment is that many young men in central Mali, in particular Fulani pastoralists, have joined various armed groups labelled 'jihadist'.

While the climate change and conflict narrative hold a prominent place in international politics and media debates, and while the Sahel is often presented as a typical example of the postulated close link between climate change and conflict (as discussed in chapter 1), empirical research on the Sahel tends to question this narrative. Instead of climate change, the main causes of conflicts in the Sahel are in general associated with state policies resulting in the marginalisation of pastoralists.

This means that while climate change and population growth may be relevant, at least in some cases, for explaining struggles over land and natural resources, political factors often provide deeper historical explanations for violent conflicts. With the current complex political and security situation in the Sahel, one may argue that today's context is taking 'revenge' on simplistic single-factors explanation (Olivier de Sardan 2021a) through the intricacies of military coups, shifting jihadist and other alliances, popular anti-colonial protests as well as the dynamics of climate change and environmental variation in relation to local land-use.

To further explore such contexts in the Sahel, this book takes inspiration from the analytical lens of 'moral economy' that may assist in more clearly seeing the values and ethics at stake among individuals deciding to join jihadist groups. In perhaps the very first contribution to the field of political ecology that emerged in the 1980s, Watts (1983) analysed moral economies in peasant communities in the Sahel (Northern Nigeria). This is political ecology in its materialist classic version rooted in neo-Marxist peasant studies from the 1960s and 70s, a canon of literature that has further inspired this book.

In a similar vein, Neumann (1998) also anchored his study in the analysis of local moral economies within the combined frameworks of political ecology and peasant studies. More specifically, he studied moral economies among local villagers who were evicted from Arusha National Park in Tanzania and whose land-use continued to be negatively impacted by the park. Neumann focused on the senses of justice among the villagers, which revolved around rights to subsistence and to customary use of land. He contrasted state laws and policies as well as the violence of park guards with the normative standard of local moral economies. The local resistance to state laws and practices was, however, largely passive and non-violent in the form of 'weapons of the weak' (Scott 1985) rather than openly violent, as currently is the case in the Sahel.

More recently, however, moral economy has largely faded out of focus within studies of political ecology of land and environments in Africa as well as in research related to land-use conflicts in the Sahel (with Turner (2004) and Nwankwo (2023) as notable exceptions).

This book represents a suggestion to revert back to such early contributions to political ecology and peasant studies using the lens of moral economy as an entry into better understanding the complexities and especially the sense of injustice driving many pastoralists in particular to join jihadist groups.

The term 'moral economy' was first coined by the English Marxist historian E.P. Thompson in his studies of peasant rebellions in England in the 18th century (Thompson 1971). His point of departure was a critique of what he saw as a mainstream historical approach of not taking the agency of common people seriously. They tend not to be seen as historical agents, he argued, and their actions are often understood as mere responses to outside stimuli. It is enough to mention a bad harvest or a down-turn in trade, 'and all requirements of historical explanation are satisfied' (p. 185), he wrote:

> Today, one could add 'drought' or 'increased temperatures' as examples of outside stimuli used as an explanation of human behaviour, as climate change is often interpreted as a driver of migration and conflict in the Sahel.

Moreover, Thompson found 'rebellions of the belly' – the idea that peasant rebellions were caused by high food prices or simply unemployment – to be paternalist. It is self-evident that people are more inclined to protest when they are hungry, but over-focussing the analysis on this aspect remains reductionistic and reflects a narrow view of 'economic man'. The interesting question is how custom, culture and reason modify social behaviour. Hence, social behaviour cannot be reduced to simple reactions to outside stimuli without context, Thompson argued.

In order to understand rebellion – and in contrast to reductionistic views – one therefore needs to understand people's motives and sense of justice within a moral economy. In the history of Europe from at least the 15th century, threats to this moral economy, usually through the privatisation of communally used land and resources, led to numerous local or more widespread peasant revolts (Thompson 1971, Federici 2004, Angus 2023).

In the peasant studies literature that emerged from the 1960s, peasants in the modern age have been seen as small-scale farmers, but also pastoralists and agro-pastoralists are seen as peasants in this literature (Chayanov 1966 [1925]). While peasants are conceived as partially integrated in markets and therefore part of the capitalist system, a peasant unit is still not a fully capitalist unit of production, because it operates according to its own complex logic involving both market and non-market incentives as drivers of land use change

(Ploeg 2013). Peasant labour, for instance, is often derived from family members rather than provided in exchange for a wage. Therefore, and linked to what Chayanov terms 'self-exploitation', peasant farms tend to be more competitive than capitalist farms. When capitalist farms go bankrupt, peasants in the same areas may work longer hours and sell at lower prices (Ploeg 2013). Peasants are also able to put to use more marginal land than capitalist farms, as they use the land more intensely, adding labour force to each unit of land, which leads to higher yields.

Peasant studies have more recently developed into 'critical agrarian studies' as a consequence of contemporary 'agrarian questions' involving more actors than peasants. Critical agrarian studies focus on how agrarian life and livelihoods shape and are shaped by the politics, economics and social worlds of modernity, while still aiming to explain the motivations and behaviour of peasants (Edelman and Wolford 2017).

In his study of peasant rebellions in south-east Asia, Scott (1976) developed the idea of moral economy further – seeing moral economy primarily as a subsistence ethic meaning that it implies a perceived right to basic subsistence. A threat to this right creates a moral indignation or anger that may lead to rebellion. Moreover, a moral economy is based on a notion of economic justice and of exploitation. Such anger is, however, not sufficient for rebellion to take place. It consists of social dynamite rather than its detonation, says Scott.

> A study of the moral economy of peasants can tell us what makes them angry and what is likely, other things being equal, to generate an explosive situation. But if anger born of exploitation were sufficient to spark a rebellion, most of the Third World (and not only the Third World) would be in flames. Whether peasants who perceive themselves to be exploited actually rebel depends on a host of intervening factors – such as alliances with other classes, the repressive capacity of dominant elites, and the social organization of the peasantry itself (Scott, 1976: 4).

While Scott (1976) did not focus on such 'intervening factors', these factors were at the core of Eric Wolf's study of six peasant wars in the 20th century (Mexico, Russia, China, Vietnam, Algeria, Cuba) (Wolf 1969). The long-held view in Marxism that peasants without outside leadership cannot make a revolution was further supported by Wolf. Peasants would therefore depend on some external actor 'to challenge the power which constrains them' (Wolf 1969: 290).

Variations of 'primitive accumulation' preceded all these rebellions studied by Wolf. This is a process famously identified by Marx in chapter 27 of Volume 1 of 'Capital' as a key component in the development of capitalism (Marx 1976). He used the example of the 'parliamentary enclosures' in northern England that mainly took place in the period 1750–1850. Before the enclosures, farmland was usually formally owned by Lords of the Manor, but with local farmers in nearby villages and hamlets having use rights in what was in practice a commons. These rights were seen as obstacles to agricultural modernisation as well as to the extraction of coal that was becoming a source of energy for the emerging industries. In addition, the principal gain from enclosures was the increased value of properties, which made it possible to charge higher rents (Mingay 1997).

It was usually the lords themselves that took the initiative to these enclosures, or privatisations, of the commons. In this 100 year period, the enclosures of about 4000 commons were enacted in Parliament. These enclosures represented about 20–25 per cent of the land area of England and Wales (Mingay 1997). There was a committee of three commissioners that was responsible for the sub-division of each commons. One commissioner represented the lord, a second represented the tithe holders (usually the Church), and a third represented the remaining owners collectively (Mingay 1997).

The result was that many villagers lost access to land for grazing or farming and ended up as a labour reserve for the new cotton textile factories that were fast popping up around the city of Manchester from the late 18th century. The loss of access to land also led to local resistance, for instance in the form of tearing down hedges and fences (Angus 2023). In addition, there was formal resistance in Parliament, and during 1730–1839 26 per cent of the Bills failed or were withdrawn (Mingay 1997).

This privatisation of commons led to the combined effect of capital accumulation for some and dispossession for others. In this classic example from England, many of those who were dispossessed became employed in the factories that emerged with the industrial revolution. Today, however, the dispossessed in the Global South tend to be 'surplus populations'. They are only in the way and no jobs are waiting for them in the manufacturing industry for instance, because they are not relevant to the current needs of capital (Li 2010). Typical surplus populations are found in prisons, in refugee camps and in ghettos. There is, however, another dynamic that Li (2010: 69) focusses on – where the 'places (or their resources) are useful, but the people are not, so that dispossession is detached from any prospect of labour absorption'.

Such places may be sites of conservation of 'wilderness' or forests, places with capital investments for agricultural large-scale development, or big afforestation projects for climate mitigation. This kind of radical land-use change through processes of primitive accumulation holds considerable potential for producing surplus populations.

To stress that primitive accumulation is not only a historical process but also one taking place today in various parts of the world, Harvey (2003) introduced the term 'accumulation by dispossession'. This term is also more explicitly highlighting the combination of dispossession of some people and capital accumulation to the benefit of others. It emanates from the idea that privatisation of commons and dispossession are products of capitalism. In political ecology, the introduction of this term has sparked a renewed interest in the combination of dispossession and capital accumulation (Büscher 2009, Corson 2011, Kelly 2011, Benjaminsen and Bryceson 2012).

Levien (2022), however, criticises Harvey for assuming that all land dispossession is driven by private capital. As an alternative, he advances the concept of 'regimes of dispossession' defining dispossession 'as a social relation of coercive redistribution mediated by those who control the means of violence in a particular territory' (Levien 2022: 42–43). His argument is that state-driven dispossession has different drivers and politics than marked-driven land loss. While the latter tends to be a piecemeal and individual process, the former can lead to events bringing large social groups into direct confrontation with the state. Hence, while primitive accumulation and accumulation by dispossession may be useful concepts and processes to study, there is a lack of focus in this approach on state-driven dispossession, which is common in the Global South, according to Levien.

This means that in addition to the politics created by capitalist dynamics and primitive accumulation, there are also state politics and land dispossession associated with elite capture that happens independently of the penetration of capitalism, as was the case with pre-capitalist privatisation of commons in Europe (Federici 2004).

In peasant studies and critical agrarian studies, there is a substantial literature linking rebellion to land issues in Latin America (Taussig 1987, Smolski and Lorenzen 2021) and to the ongoing Naxalite rebellion in India for instance (Gupta 2007, Parashar 2019). For some reason, however, there is little scholarly work on this link in Sub-Saharan Africa. Land dispossession generally falls outside the scope and focus of academic debates on the rise of jihadist groups in the Sahel. There is, however, a broader West African literature on the drivers of recruitment into rebel (rather than jihadist) groups pointing at agrarian and land-related causes (Richards 2004, Cavanagh 2018).

MORAL ECONOMY AND JIHADISM IN THE SAHEL

Sahelian pastoralists have, over a long span, complained about a continued loss of access to dry season pastures and livestock corridors, which has been facilitated by a general negative policy discourse surrounding Sahelian pastoralism among governments and aid organisations (Thébaud and Batterbury 2001, Pedersen and Benjaminsen 2008).

State policies that favour settled agriculture at the expense of mobile and flexible livestock production undermine not only pastoralists' access to land but also livestock-keeping as an economic activity in general. Pastoral marginalisation was also one of the main drivers of the Tuareg rebellion that triggered

Mali's civil war in the 1990s (Benjaminsen 2008). Many Tuareg and nomads in general had for a long time felt they were being marginalised by state policies of modernisation.

These grievances therefore emerge from moral indignations about economic injustice and exploitation related to the governance of land and natural resources.

As mentioned in chapter 2, in 1982 Mali agreed on the implementation of a Structural Adjustment Programme. This led to a reduction in government spending and in the number of civil servants in all sectors except in forest governance where the number of forest officials in rural areas throughout the country increased many-fold from the mid-1980s. This increase was a direct result of the new global attention to and support of sustainable development and in particular the need to fight desertification. The increased presence of forest agents to police village and pastoral land was seen by the government and international aid donors as necessary to stop an advancing desert in effective ways. This meant that the Forest Service received international support from the World Bank and other donors.

The result was a top-down approach based on handing out fines to people for activities that were seen as leading to environmental degradation and desertification. Through these practices, forest agents became a constant threat to rural people and their livelihoods all over the country, leading people not to see them as defenders of nature, but as money-grabbers.

The senses of justice leading to this rebellion are further discussed in the section on the biography of the two jihadist leaders below. In brief, the argument is as follows – the Tuareg and Fulani grievances and associated senses of justice have different origins, although for both groups a defence of pastoralism as a viable livelihood has been in the driving seat of the uprising. From 2012, when the uprising became 'jihadist', and when the Fulani joined, it appealed as well to the subordinate classes (Bella and Rimaïbé – the former 'slaves' among Tuareg and Fulani respectively) who saw this as an opportunity for social liberation. In this way, the current 'jihadist' revolt in Mali is not so much a sign of radicalisation of Islam as it is a sign of islamization of a radical revolt (Holder 2023).

The colonial history of Kidal and the Ifoghas clan plays a key role in understanding the Tuareg rebellions of 1963 and 1990 and thereby also the development of the current crisis in Mali. In addition, the fact that the Ifoghas felt betrayed by France in 1960 when the former colonial power decided not to support the establishment of an independent Tuareg state in northern Mali, led the Tuareg in Kidal to be in opposition to the Malian state from the outset. The subsequent paternalist policy of the Keita regime towards the pastoral way of life further increased the will to carry out a rebellion.

The rebellion of 1963 was, however, quickly crushed by the Malian army using fighter bombers and public executions. This further led many young Tuareg to want to take revenge, and a new revolt was in the planning from the 1970s in Algeria.

In the Fulani case, after Hamadoun Kouffa had established Katiba Macina in 2015, he distributed a number of speeches on WhatsApp. By emphasising a social interpretation of Islam, he repeatedly spoke about social justice and social equality, in particular addressed to the subordinate Fulani (Rimaïbé) (Holder 2023). Hence, the jihadist discourse has also strengthened an 'awakening' among the Rimaïbé.

In his speeches, Koufa also refers to the Fulani struggle with the Bambara Empire based in Ségou, which led to the emergence of the Macina Empire. This was a Fulani-based theocratic Islamic state that ruled central Mali from 1818 to 1862 (Johnson 1976). It prohibited singing and dancing, and women were secluded. The capital, Hamdalaye:

> was policed to enforce public morality, and an overseer of public morals was appointed in other towns – one of his duties was to oversee weights and measures and the quality of goods sold on the market. He was doubtless charged also with the enforcement of controlled prices of foodstuffs (Johnson 1976: 484).

The state was maintained through a complex system of taxation that provided funds for the bureaucracy, the mosques, and the armed forces, in addition to 'support of widows and orphans, the aged, insolvent debtors, pilgrims and other travellers' (Johnson 1976: 483). This historical knowledge, which is shared by many in Mali, gives an insight into the moral economy that Koufa in his speeches says he wants to reinstate.

From 2018, when army attacks on the Fulani population intensified including widespread human rights violations and summary executions, Koufa's rhetoric started to change from a dominant focus on social justice to also include a defence of the Fulani including of pastoralism as it is a central part of Fulani identity. The rhetoric also changed to attacks on 'the Bambara state' referring back to the Macina Empire, the golden age of Fulani power, and its jihad against Bambara towards the south (Holder 2023). The view of the Malian state as a Bambara state that does not understand pastoralism as a livelihood strategy is something Koufa has in common with many Tuareg.

This struggle is both discursive and material. In a material sense, both the pastoral moral economies of the Tuareg and the Fulani are grounded in the control over pastures and other natural resources. The associated grievances therefore emerge from moral indignations about economic injustice and exploitation related to the governance of these resources.

We will, in the following, discuss the material grievances of two such moral economies – related to forest governance and land-use conflicts. These are key issues repeatedly raised in numerous interviews and conversations with people from central and northern Mali during the last three decades.

After the Tamanrasset Peace Agreement in January 1991 between the government of Mali and the Tuareg rebel groups in the north that had taken up arms in June 1990, public meetings were held in the northern regions to explain the content of the agreement to communities. A large number of interventions at these meetings concerned grievances related to the practices of the Forest Service. Furthermore, with the transition to democracy from March 1991, newly established newspapers wrote extensively about the harassment by foresters in rural areas, and at the National Conference in Bamako in June–July 1992, this was one of the returning themes taken up by representatives from various political and rural organisations (Benjaminsen 1997).

More than two decades later, Katiba Macina has taken control over the inland delta area in central Mali (Figure 3.1). The general corruption of state officials and politicians in the delta area is one of the key recurring themes in Koufa's speeches on WhatsApp (Thiam 2017). In addition, he often points out the Forest Service as a particular problem, knowing well that its practices have been a major concern in rural areas.

In one speech, Koufa states for instance:

> Our fight is not directed against the peaceful populations who are victims of bad governance by the state including a bad administration and a corrupt justice system. We can also add the behaviour of the judges and the agents of the Forest Service who are condemned in the whole region. These agents are condemned by local populations because they are recognised predators (translation from Fulfulde via French).

In this way, Koufa demonstrates how to play on people's moral indignation and sense of injustice and exploitation caused by the predatory behaviour of political and economic elites in Mali. Such a discourse is much more efficient to attract adherents to the organisation rather than one based purely on religious arguments.

The dispossession of people's access to forest resources in Mali since especially the 1980s is also more in line with Levien's 'regimes of dispossession' based on the control over the means of violence. While Mali has been integrated into an international capitalist system, although in its periphery, capitalist penetration is less evident in the pastoral sector in northern and central Mali still dominated by moral economies. On the other hand, the forest sector has clearly been under the influence of a global (neoliberal) environmental management discourse for several decades (Adger et al. 2001), and in southern Mali and peri-urban areas, a neoliberal reform of the forest sector took place

Source: Author's own.

Figure 3.1 *Map of Mali*

from 1996 initiated by the World Bank (Gautier et al. 2013). The purpose was to establish fuelwood markets in the peripheries of urban areas and enclosing the rights to harvest and to sell wood. These measures were seen as necessary to fight the fuelwood crisis. However, the reform led to adverse effects such as more confusion around forest rights and increased elite capture.

In the agricultural sector, especially in high-potential areas such as Office de Niger and other parts of the delta, there have been a number of international investments (Touré 2022) leading to more obvious forms of accumulation by dispossession through the grabbing of village land at the benefit of investors.

There has also been widespread moral indignation among rural people in central Mali and, especially among pastoralists, related to land-use conflicts and their lack of resolution. In particular, the inland delta of the Niger river has been marked by intensive land-use conflicts during the last few decades (Moorhead 1991, Turner 1992, Barrière and Barrière 2002, Cotula and Cissé 2006, Ba 2008, Benjaminsen and Ba 2009, Benjaminsen et al. 2012). The most violent of these conflict types tend to involve the encroachment of farming on pastoral land, especially livestock corridors and dry season pastures. This can be encroachment from small-scale village farming, but often also from large and medium-scale rice irrigation schemes.

The main fodder resource in the delta is burgu (Figure 3.2), which grows on deeper water than rice, which again has expanded rapidly at the expense of burgu during the last few decades. According to Kouyaté (2006), about one quarter of the burgu areas in the delta had at the time been converted to rice fields since the 1950s.

There is a long-standing pastoral grievance related to the blocking of live-stock corridors and loss of burgu pastures. Each year, several million cattle enter the delta at agreed-upon dates in late November and at specific entry points. The current system of managing burgu in the delta was established and formalised by the Macina Empire. The delta was at the time divided into 37 territories (*leyde*) where a pastoral chief (*jowro*) in each leyde was given the responsibility to manage the burgu pastures. This system included establish-ing livestock corridors (*burtol*) from the dryland areas used during the rainy season and into the delta, as well as defining the entry points (Benjaminsen and Ba 2009).

From the Macina Empire and into the colonial period, pastoralists paid only symbolic fees in kind to the jowro at the entry point, but during the last few decades the fees herders have had to pay per head of livestock have continued to increase, as politicians and government officials started to see these events as new sources of income. As the jowros depended on the support of the state to be able to continue to manage the burgu pastures, some jowros distributed all fees collected at these occasions just to keep the powerful people on their side (Turner 2006, Benjaminsen and Ba 2009). By paying these bribes, the jowros established indispensable relations with government representatives.

Needless to say, increasing grazing fees were disliked by pastoralists, espe-cially those originating from villages in the drylands of Sénou, Hayre and other areas around the delta who have to pay higher fees than herders from villages inside the delta who often do not pay much at all. With Katiba Macina taking control over the delta from 2015, grazing fees to burgu pastures were abolished, which was obviously a popular move among dryland-based pasto-ralists. However, in 2018, the jowros complained to the Shoura (the committee of leaders) of Katiba Macina and asked for permission to start collecting fees again, which was granted, although the rates were reduced compared to the past.

Source: Abdoulaye Sow.

Figure 3.2   Burgu pastures in the inland delta of the Niger river

Many dryland pastoralists rejected, however, to comply with this decision, arguing that the holy Quran states that pastures are a free gift from Allah. They therefore asked for a reversal of the decision, but the Shoura still decided to keep the fees. This resulted in dryland pastoralists leaving Katiba Macina to join Dawlat il Islamia, which is a group established in late 2019 that has become part of Islamic State in the Greater Sahara.

Moreover, during fieldwork in the delta in the years 2006–09 people interviewed repeatedly complained about having to pay bribes to the local administration. It was often stressed in these conversations that the one who pays the most to the administration or to the courts is also the one most listened to. By receiving payments from both parties, decisions on land-use conflicts in the courts become ambiguous, which again contributes to keeping conflicts alive. In conversations and interviews, we often heard complaints that the rural population has become milking cows for government employees and political elites (Benjaminsen and Ba 2009).

The expansion of rice fields at the expense of burgu has to a large extent been driven by the para-statal company Office Riz Mopti (ORM), which was established with World Bank funding in 1972 as one of several sector-orientated state development agencies. ORM was mainly funded by the World Bank until 1991 and thereafter by the African Development Bank.

Its main focus has been to expand the area under controlled flooding for rice cultivation in the delta. This means constructing a series of dikes with weirs to control the water level and canals to distribute the water to fields. The irrigated area is divided into plots of one hectare each and all Malians can apply for plots, although people with access to information and political connections have had an advantage leading to the establishment of a new group of urban landowners in the delta, at the expense of pastoralists in particular. This process has led to communal burgu pastures being converted into private rice fields.

According to ORM officials interviewed in 2006, ORM never took livestock problems into account. This was rather the task of another para-statal – Opération de Développement de l'Elevage dans la région de Mopti (ODEM) – which was established in 1975 to promote livestock development, also with World Bank funding. The aims of ODEM were to restore ecological equilibrium in the pastoral sector based on the idea that there was widespread overgrazing, increase the productivity of livestock (assuming that the pastoral system in central Mali was unproductive and inefficient), improve the socio-economic conditions of the population, and promote the marketing and export of cattle to relieve pressure on grazing land and to earn foreign exchange. However, the approach followed was technocratic and top-down leading to disappointing results (Shanmugaratnam et al. 1991, de Bruijn and van Dijk

1995). In 1991, the World Bank discontinued its funding, and the activities were phased out.

Before closing down activities ODEM had, however, contributed to an increase land conflicts through drilling new wells in pastoral dryland areas. The new water sources attracted farmers to settle and conflicts over land emerged (De Brujn and van Dijk 1995, Benjaminsen and Ba 2021, see chapter 4).

People interviewed in the delta in 2006 also stressed that neither farmers nor pastoralists were consulted when ORM confiscated burgu land and closed livestock corridors. One jowro said that ORM had robbed his community of land worth many billions of FCFA. He called it 'an aggressive hold-up' and added that 'ORM – it's a disaster'. Another jowro added that 'everything ORM has done is bad'.

Corruption by a rent-seeking bureaucracy has been increasing rather than reducing land-use conflicts. For instance, in order to re-open livestock corridors, jowros interviewed said they had to pay off a number of state technicians and administrators. Another well-informed interviewee sitting in a key political position stressed to us that the administration benefits from the jowros only having informal or customary power without formal statutory rights. As long as they only have informal power, they need the support of the administration in order to be able to manage pastures effectively. This support is obtained through bribes. 'A jowro who tries to be correct without paying off the administration will never be able to do anything', the interviewee added (Benjaminsen and Ba 2009).

This also goes for the legal system. Both sides in land-use conflicts that are dealt with by the courts have had to spend large sums of money to bribe the judges and their entourage. Through receiving payments from both parties, the courts' decisions have tended to become ambiguous, which again has contributed to perpetuating conflicts (Benjaminsen and Ba 2009).

The loss of burgu pastures in the delta over the last few decades in tandem with the blocking of many historical livestock corridors is at the core of pastoral grievances associated with a common feeling of moral indignation and injustice in the delta in central Mali. There is also a sense of historical loss related to Fulani identity in central Mali, which is closely linked to the heritage from the Macina Empire. Such feelings among many Fulani are also aggravated by the sense that these processes degrade the pastoral system inherited from the Macina Empire. Essential parts of this heritage, communal burgu pastures managed through specific rules and livestock corridors established by the Macina Empire, have been encroached upon by what is seen as corrupt forms of 'modernisation' during the last few decades.

The fact that these encroachments have been part of a government policy supported by foreign aid donors that have benefitted government officials and urban elites through corruption and resource grabbing, has further boosted these feelings and a general grudge against the government that the jihadist leaders have later been able to make full use of to their advantage.

The dispossession in this case cannot be seen as a product of a regime of dispossession (Levien 2022) based on controlling the means of violence. The penetration of capitalism into this periphery has, however, played a role, as have the effects of international development aid. The Office Riz Mopti project funded by the World Bank and the African Development Bank has led to the privatisation of land that used to be communal burgu pastures under jowro control. This land is currently de facto private rice fields, which is one of several examples of conversion of burgu pastures into private agricultural land.

But still the case does not entirely fit the idea of accumulation by dispossession either. There is more than the penetration of capitalism that is behind the processes of dispossession and accumulation. Perhaps, it represents a third type – neopatrimonial dispossession associated with 'belly politics' (Bayart 1993) seen as 'a complex mode of government' that denotes 'the accumulation of wealth through the tenure of political power' (Bayart et al. 1999: 8). This has also been referred to as the neopatrimonial state 'in which officeholders systematically appropriate public resources for their own uses and political authority is largely based on clientelist practices, including patronage, various forms of rent-seeking, and prebendalism' (van de Walle 2001: 52).

This neopatrimonialism may, however, be partly explained by colonialism. Olivier de Sardan (2021b) argues that the Sahelian states were brutally created basically from nothing by colonial powers. These Sahelian states became 'despotic in nature' and far from the model of the French state, and they were continued after independence developing their own bureaucratic culture and practice of elite capture. The combination of top-down governance, bureaucratic corruption, and lack of provision of services has over time led to a lack of trust in the state among its citizens.

THE ROLE OF JIHADIST LEADERS

Corruption and embezzlement of aid relief funds destined to the north was in fact one of the justifications that Tuareg rebels used to start the revolt in 1990 (Benjaminsen 2008). The leader of this rebellion was Iyad Ag Ghaly who is currently also the leader of Ansar Dine and seen as the paramount jihadist leader in Mali.

This final section of this chapter discusses the trajectory of Ag Ghaly and the leader of Katiba Macina, Hamadoun Koufa, and how these two jihadist leaders have been able to harness the frustrations towards the state especially among pastoralists in central and northern Mali.

In 2011, Iyad Ag Ghaly founded Ansar Dine after the separatist Tuareg movement MNLA (Mouvement National de Libération de l'Azawad) had refused to name him its head (Eizenga and Williams 2020). He was born in 1958 and comes from Boghassa village close to the Algerian border in the Kidal region in the far north of Mali, and belongs to the Ifoghas clan of the Kel Adagh Tuareg. The 'Tuareg problem' in Mali is generally seen to have originated in Kidal among the Kel Adagh (meaning people of Adagh referring to the Adagh mountains north of Kidal town).

The Ifoghas and Idnan are the two most powerful clans among the Kel Adagh. Before colonisation, there was a certain power balance between these two clans, but the French colonial government chose to give preference to the Ifoghas whose leader was made Amenokal (supreme chief) (Lecocq 2010). In the pre-colonial time, the Kel Adagh had tended to be dominated by neighbouring Tuareg groups such as Kel Ahaggar and Ouelleminden, and they therefore tend to look back at the colonial period as a good period, since the French ended this domination. They therefore saw the French as allies and also fought with the colonial power against other Tuareg groups that resisted occupation. It was, however, seen as treason when France left the area in 1960 and transferred power to politicians in the south, because the Kel Adagh thought they had been promised independence (Lecocq 2010).

The Kel Adagh were therefore in opposition to the Malian state from its inception, and the first revolt, inspired by the FNL (the national liberation front) anti-colonial war in neighbouring Algeria, broke out in 1963 (Lecocq 2010).

In addition, the first Malian President, Modibo Keita, was inspired by ideas about industrialisation and agricultural modernisation, and nomadism was seen as an obstacle to such development. Sedentarisation of nomads and converting them to 'productive' citizens by taking up farming became a key component of the government's approach in the north (Benjaminsen and Berge 2004a). This meant that the nomadic way of life was considered backward, unproductive and undesirable. The pastures in northern Mali were even labelled 'le Mali inutile' ('the useless Mali').

The revolt of 1963 was, however, severely crushed through the use of fighter-bombers and public executions, which led many Tuareg to migrate to Algeria and Libya (Ag Baye 1993). Ag Ghaly was, however, part of a second wave of migration to Libya that took place after the drought of 1973–74. Khadafy was keen to attract young Tuareg to train them in special camps and use them in Libya's various military operations.

In 1979, Ag Ghaly had his first war experience when he was sent to fight for the Palestinian cause in southern Lebanon together with 660 other Tuareg. They fought in Lebanon until the Israeli invasion in 1982. Back in Libya, the Tuareg soldiers were given the choice of staying in the army or leaving the country. There were no longer special camps for Tuareg fighters as had been the case in the 1970s. However, in March 1986, Khadafy established a new Tuareg military camp – this time to support the Libyan offensive in Chad from July 1986 until the cease-fire in September 1987 (Boiley 1999).

Already in 1982, Tuareg fighters from northern Mali had created a clandestine resistance movement in the camps in Libya with Ag Ghaly as its elected leader. He was a recognised soldier and had also already made a political impression through talks with Libyan authorities on behalf of Tuareg soldiers (Boiley 1999).

He had also witnessed how the international community, including Arab nations, did not stand up for the Palestinians, and had concluded that the Tuareg could not expect any outside support. Hence, they needed to only trust themselves and to create their own revolution. The consequence of this insight was to abandon any open political opposition in Mali, which anyway did not have much prospect within Mali's military regime, and turn towards armed rebellion. In a meeting in Tripoli in 1987, the resistance movement was formalised with a council of about 30 members and with Ag Ghaly as its secretary general. A charter with objectives was formulated. The main objective was liberation of the area in northern Mali that they named Azawad, and to create an independent state with that name (Boiley 1999).

Thereafter, Tuareg fighters started to return to Mali, but very gradually in order not to raise any suspicion. Cells were created with specific defined tasks in the forthcoming uprising. The first attack took place in Menaka in late June 1990 and was led by Ag Ghaly; it was only equipped with some camels, two hunting rifles, five Kalashnikovs and some swords (according to Ag Ghaly interviewed by Boiley (1999)). Menaka was chosen as a target because they knew there were weapons and NGO vehicles. This first attack in this rebellion led to the killing of 36 people, while only one Tuareg fighter was injured. The rebels got away with 12 vehicles from the NGO World Vision as well as a large amount of arms, diesel and food (Boiley 1999).

Initially, there was, however, a lack of support for the rebellion among Tuareg outside Kel Adagh. This quickly changed. The attack unleashed a campaign of indiscriminate violence from the Malian army against nomads. Civilians who had never even heard of any rebels were massacred (Poulton and Ag Youssouf 1998). 'Within a few weeks, the Malian army had created hundreds of new "rebels", as Tuareg youths fled into the hills to escape massacre' (Poulton and Ag Youssouf 1998: 56). In addition, the army made no difference between Tuareg and Arabic Moors, turning these latter also into rebels. Hence, the 'Tuareg rebellion' was not only Tuareg anymore, but had been turned into a nomadic or pastoral revolt.

Simultaneously with the escalation of the conflict in northern Mali, the democracy movement in Bamako was gaining force. Realising that a rapid victory in the north was not within reach and that he might need the troops in Bamako, the President at the time, Moussa Traoré, initiated peace negotiations. Towards the end of 1990, direct talks between the government and rebel leaders were initiated resulting in the signing of the Tamanrasset Peace Treaty on 6 January 1991. The key point in this agreement was to give northern Mali a 'special status', which implied that 47.3 per cent of the national budget should be allocated to the north. But this also meant that separation from Mali was not on the table. Some hardliners within the rebel movement were not happy about this, but Ag Ghaly, who had signed the treaty on behalf of the rebels, was among the moderates who could accept to stay within Mali, only with more autonomy (Lecocq 2010).

After massive pro-democracy demonstrations in Bamako in March 1991 culminating in the coup d'etat against the Traoré regime, an interim government led by Amadou Toumani Touré (ATT) organised parliamentary and presidential elections in 1992. As part of this wave of popular democracy, a national conference with more than 1000 participants was held in Bamako in June and July of 1992. In the conference it was generally recognised that the rebellion had played a role in bringing down the military regime. But at the same time, the Tamanrasset Treaty was not recognised, because it was seen as giving away too much to the Tuareg. Ag Ghaly, as the leader of the rebellion, pleaded, however, at the conference for more federalism and regional political and economic autonomy (Lecocq 2010).

However, the violence did not stop despite the democratisation process, the Tamanrasset Treaty, and a National Pact that had been signed in April 1992. The army seemed to continue to operate on its own in the far north, outside the control of the newly elected Konaré government in Bamako, which led to more violence against civilians and many people fleeing to neighbouring countries. The response was more armed attacks by the rebels on the army and government institutions and their representatives leading to retaliations on civilian Tuareg and Moor and their properties, especially in the urban areas (Benjaminsen 2008).

In late 1994, the government reshuffled the armed forces including appointing a new minister of defence and withdrawing military units that were associated with bad discipline and excessive use of force. This move seems to have helped President Konaré to gain control of the army. From November 1994 to June 1995, a number of public meetings were held in the north initiated by local leaders or by rebel groups. These discussions calmed down the situation and prepared the ground for inter-community meetings facilitated through assistance from the UNDP, the Norwegian government and a group of Malian facilitators coordinated by Kåre Lode from the Norwegian Church Aid (Lode 1997, Poulton and Ag Youssouf 1998). During August 1995 to March 1996, 37 inter-community meetings were held. Interestingly, these discussions totally sidestepped both rebel leaders as well as the army and top government officials creating conditions for peace at the grassroots. The whole process culminated in the 'Flamme de la Paix' ceremony in Timbuktu on 27 March 1996 where 3000 hand weapons collected from the rebels were burnt (Benjaminsen 2008).

Following this Peace Agreement between the government and the Tuareg rebels, Ag Ghaly normalised relations with the government who used him to help negotiate the release of Westerners kidnapped by AQMI in the early 2000s, such as 14 German tourists in 2003. In 2008, he was appointed by President ATT as a cultural counsellor to the Malian consulate in Jedda in Saudi Arabia, perhaps to create a distance between him and Malian politics. In 2010, Ag Ghaly was, however, expelled from Saudi Arabia because he had apparently established closer contacts with radical jihadists in the country than what the authorities appreciated.

Prior to his intermezzo in Saudi Arabia, in 2005, Ag Ghaly had made an unsuccessful attempt at becoming Amenokal of the Kel Adagh. The next year he joined three Tuareg officers, who had been integrated into the army and then defected, in a new limited rebellion based in Kidal. Their demand was a particular status for the Kidal region after having attacked the Malian military positions in Kidal and Menaka, before withdrawing to the Adagh mountains. Negotiations were immediately initiated by ATT with Algerian facilitation leading to the Algier agreement of 4 July 2006.

Ag Ghaly and the other leaders of this rebellion had stressed that they were not involved in any jihadist agenda, that their fight was merely for better living conditions in Kidal, and that they had no links with the GSPC Salafist group that had infiltrated northern Mali from Algeria after the civil war there (Lecocq 2010).

Some interviewees,[1] who knew Ag Ghaly at the time, state, however, that he was under the influence of AQMI already from 2004. To increase their influence in northern Mali, AQMI leaders inter-married in Arab communities west of Timbuktu, in the Lake Faguibine area, and worked to infiltrate communities and social networks in the north. In addition, they aimed in particular to include community leaders in their cause, such as Ag Ghaly and Hamadoun Koufa in central Mali.

Koufa's integration into jihadist networks in northern Mali was facilitated by the fact that Ag Ghaly and Koufa already knew and respected each other, and that Koufa was already a respected Islamic scholar and preacher. Hence, Ag Ghaly was supported by AQMI to create a *katiba* (Ansar Dine) in 2011, while Koufa created his Katiba Macina in 2015. These two jihadist groups have since 2015 controlled large parts of northern and central Mali with Ag Ghaly as the chief of JNIM (Group to support Islam and Muslims), which is linked to Al Qaeda and that in addition to Ansar Dine and Katiba Macina includes Al-Mourabitoun and AQMI.

Hamadoun Koufa was born in 1961 in Nianfunké in the Mopti region as Hamadoun Sangaré. From the 1980s, he became known in the region for his eloquence and poetry. He travelled around the region to preach Islamic virtues, and in the early 2000s, he also spread his message through a local radio station, which boosted his regional fame. Recorded cassettes also started to circulate increasing his reputation. In his preaching, he increasingly attacked religious leaders, traditional elites and the state administration. His critique of other Islamic scholars in the region led to replies and counter-attacks, which may have contributed to his radicalisation (Thiam 2017).

In his critique of other Islamic scholars, he argued that instead of asking brave peasants for *zakat* (tax), they should themselves work the land like these people do. He also criticised the widespread practice of sending out kids (*talibé*) to beg. For him, living off the work of kids (begging) is more sinful than drinking alcohol (Cissé 2018).

In 2004, he visited the 'Jamaat Tabligh' society in Pakistan, which has been accused of having links to Islamic terrorist groups. Upon his return to Mali, he obtained funding from Pakistan to rehabilitate a mosque in Sinakoro, 20km from Mopti town. The debate about a new proposed Family Code in Mali in 2009, giving women a stronger position in society with, for instance rights to divorce and inheritance, created an outcry from conservative Muslims including Koufa, and may have further contributed to his radicalisation (Thiam 2017).

As already mentioned, Koufa's strong position among Fulani pastoralists must also be understood in relation to his continued references to the Macina Empire in the 19th century, which is seen as the golden age of Fulani power at a time when a Fulani jihad was in conflict with a Bambara state further south based in Ségou. Koufa's rhetoric about the Malian state as a Bambara state that does not understand pastoralism is something he has in common with many Tuareg.

Following the rebel and jihadist take-over of northern Mali in late 2011 and early 2012, Koufa travelled to Timbuktu in July 2012 where he received military training before he joined Ansar Dine as a close collaborator of Ag Ghaly, who also became a mentor to him. In this period, he also participated in the planning of the attack on Konna that took place in January 2013.

In sum, Ag Ghaly and Koufa have personally played decisive roles in rallying Tuareg and Fulani, respectively, to join Ansar Dine and Katiba Macina. Their positions of power and authority have been achieved through a combination of their personal histories, knowledge of how to draw on local grievances, and international influences, inspirations and sometimes support from Libya, Algeria, Saudi Arabia and Pakistan.

MATERIALIST POLITICAL ECOLOGY AND THE SAHELIAN CRISIS

Scholarly analysis of the current Sahelian crisis tends to be dominated by perspectives from political science, international relations and studies of religion. This literature has generally focused on how global political economic developments and Islamist thinking and organisation relate to national Sahelian dynamics. It has, for instance, discussed the largely overlapping topics of international jihadism and its tributaries in the Sahel (Huckabey 2013, de Castelli 2014, Harmon 2014, Lounnas 2014, Walther and Christopoulos 2015), drug trafficking and hijacking of hostages as sources of funding for armed groups (Daniel 2012, Lacher 2013, Detzi and Winkleman 2016), the dynamics, politics and history of the Tuareg rebellions in relation to the crisis (Cline 2013, Zounmenou 2013, Bøås and Torheim 2013, Bøås 2015), problems in the national democratic system and the weakening of the state (Gonin et al. 2013, Baudais 2015, Ba 2016), the international military intervention (Cristiani and Fabiani 2013, Galy 2013, Hanne 2014, Ping 2014, Boeke and Tisseron 2014, Boeke and Schuurman 2015, Wing 2016), the complex and changing Islamic national landscape in Mali (Soares 2013), and the Malian crisis as a fallout from the Libya conflict (Shaw 2013, Solomon 2013).

While such contributions give a rich background to the current political crisis in Mali and other parts of the Sahel, the argument in this book is that a focus on the local political context and, in particular, the political ecological context, is necessary in order to more fully explain the expansion of jihadism. This local political ecological context is, however, generally neglected in the large and rapidly increasing scholarly literature on jihadism in the Sahel.

There are, however, some contributions that stress the role of local context more broadly. Ibrahim (2017: 10) points out that 'ideology is a necessary but not a sufficient condition for the development of a jihadist insurgency' and that 'African jihadist movements are first and foremost local movements that arise from local social and political dynamics, and their struggle is primarily geared toward addressing local – not global – grievances' (p. 8). Dowd and Raleigh (2013: 498) argue that 'violent Islamist groups emerge in and are shaped by distinct domestic contexts and issues, a feature that is obscured by a totalising narrative of global Islamic terrorism', and Dowd (2015) concludes that grievances and political marginalisation drive the emergence of jihadist groups in Kenya, Nigeria and Mali. Lecocq et al. (2013) also stress the importance of understanding the local context to explain the Malian crisis including how corruption is enmeshed in the administrative and political system, which they argue are often missed by outside observers.

In a similar vein, other researchers have also pointed to links between rent-seeking and the rise of jihadism in Mali, and suggested that grievances rather than religious conviction motivate Malian pastoralists to join jihadist groups (De Bruijn and Both 2017, Jourde et al. 2019).

Mixing such a local with a global approach, Marret (2008) calls Al-Qaeda in the Islamic Maghreb (AQIM) a 'glocal' organisation, stressing its mix of local and global interests and practices, while Boeke (2016) stresses that AQIM is both a terrorist and an insurgent organisation rooted in a global radical jihadist ideology as well as in local and national grievances. This dichotomy is also pointed out more generally by Kalyvas (2003), who argued that often civil wars are driven by an interaction between political aims and private grievances.

In the current academic debate about jihadism in the Sahel, one point of disagreement has indeed been on where to put the weight between local context and the role of global jihadist networks in explaining the growth of violent resistance (Zenn 2017, Thurston 2017, Higazi et al. 2018, Hansen 2019). However, the role played by struggles over land and natural resources is usually left out of these debates.

Shifting the attention to the political ecology of land and natural resources governance implies studying peasant motivations in relation to the politics of natural resource access as well as how authority and power relations impact on this access. While acknowledging the 'glocal' dynamics, the argument here is that the material foundation of the current crisis, including various aspects of land dispossession, are generally missing from current academic as well as policy debates.

Political ecology is a rapidly growing and diverse body of literature with a common focus on how power manifests in both discursive and material struggles over land and the environment (Forsyth 2003, Robbins 2012, Benjaminsen and Svarstad 2021). Important influences on political ecology have been neo-Marxist theory (Watts 1983), actor-orientated perspectives (Blaikie and Brookfield 1987), poststructural theory (Peet and Watts 1996), and science and technology studies (Goldman et al. 2011). Political ecology studies tend to focus on environmental subjects such as smallholders, who depend on environmental resources and who struggle to maintain both their identities and livelihoods in the face of state, civil society or corporate interventions.

While political ecology today has expanded in many directions including themes such as decolonisation, the more-than-human, climate justice, race and racism, degrowth and ontologies and pluriverses, few studies take inspiration from the field's roots at the interface with peasant studies with a focus on moral economy and land dispossession.

Drawing on such an approach means studying violence in its wider political-economic and historical context with a focus on access and control over land

and natural resources. In studies of the materiality of natural resources governance and politics, the dialectic of actors and land represents the main object
of study. This type of materialist political ecology moves, however, beyond
studying conflicts as simple causal chains with resource scarcity having negative consequences for livelihoods and again leading to migration and conflicts
(Peluso and Watts 2001).

As a pioneer in studies of the political ecology of farmer–herder conflicts,
Bassett (1988) saw these conflicts as 'responses in context', while Turner (2004)
also used political ecology in his critique of the scarcity narrative that widely
informs perceptions of farmer–herder conflicts in the Sahel, and stressed that
these conflicts should be understood as more than just resource conflicts.

In a similar vein, the insurgency in Mali is concluded to be rooted in a
threat to pastoralism in particular and its basis of subsistence. This has created
a moral indignation that has existed for decades, but which was released by a
series of international events and circumstances (Tuareg trained as soldiers in
Libya, spill-over of jihadists into northern Mali from the Algerian civil war,
influence and support from Islamist groups in Pakistan, Ag Ghaly's connections with radical jihadists during his two-year stay in Saudi Arabia, and the
possibility to connect to global jihadist networks).

Pastoralists and small-scale farmers interviewed before the crisis in 2012
complained frequently about resource dispossession caused by the heavy-
handed methods of the Forest Service as well as the pastoral dispossession of
key pastures and livestock corridors and corrupt practices in the adjudication
of land-use conflicts by local authorities and the courts.

Land and resource dispossession tends to be discussed as either market-
driven, as in accumulation by dispossession, or driven by state violence, as in
Levien's regimes of dispossession.

While the violence and dispossession in the case of forest governance discussed here has been more in line with regimes of dispossession, the conversion of communal burgu pastures to private rice fields is clearly a form
of accumulation by dispossession resulting from capitalist penetration into a
periphery. But there has also been more going on. The rent-seeking and elite
capture related to the land conversion can be seen as resulting in a form of
neopatrimonial dispossession associated with belly politics.

These widespread processes of dispossession in central and northern Mali
created a moral economic anger against rent-seeking elites that provided the
foundation of the jihadist uprising. To detonate this anger, Iyad Ag Ghaly
and Hamadoun Koufa have played key roles as jihadist leaders in mobilis-
ing popular support emerging from local grievances, while drawing on social
justice-based Islamic discourse and capitalising on external support. At the
same time, the colonial history of Kidal and the Ifoghas clan plays a key role

in understanding the Tuareg rebellions of 1963 and 1990 and thereby also the development of the current crisis in Mali.

More recently, when the uprising became 'jihadist' from 2012, and when the Fulani started to join, the uprising became also attractive to the subordinate classes who saw the rebellion as an opportunity for social liberation. Koufa's speeches about social justice have strengthened an 'awakening' among the Rimaïbé. Simultaneously, his frequent references to the Macina Empire as the golden period of Fulani pastoral power contribute to a narrative about pastoral resistance to a Bambara-dominated state that does not understand or appreciate pastoralism.

The next chapter presents a further example of how a much-debated conflict in the Sahel can be understood by employing a materialist political ecology approach.

NOTE

1. Who were interviewed by Malian colleague Boubacar Ba.

4. Materialist political ecology of a farmer–herder conflict

POLITICS OF SCARCITY

The Biblical story of the conflict between Cain and Abel, which led the former to kill the latter, is the archetypical example of what is often seen as an eternal tension between sedentary farmers and migrating pastoralists. At a practical level, this is a physical conflict over material resources, but at a general level such conflicts also represent differing worldviews involving contradicting ideas about property rights to land, economic productivity, what landscapes should look like, and also what a good life implies.

As discussed in chapter 3, most state governments tend to be dominated by a sedentary 'farmer's view' of the world. This is one reason why migrating pastoralists often finds themselves at the sharp end of state governance. The nomadic way of life is simply seen to be in the way and an obstacle to societal development.

As we also saw in chapter 3, this marginalisation, including an associated feeling of injustice, may under certain conditions lead to resistance and armed pastoral rebellion against the state. In the Sahel, such resistance has also evolved into various farmer–herder conflicts[1] as farmers are often supported by the state or even used by the state as part of its counter-insurgency.

This context is, however, often neglected in general policy or media debates about farmer–herder conflicts in Africa that tend to present such struggles as typical examples of resource conflicts associated with Malthusian processes driven by scarcity of natural resources. In the scholarly literature, this focus on scarcity as a key issue has been promoted in particular by the Environmental Security School (Baechler 1998, Homer-Dixon 1999, Kahl 2006). This School sees 'environmental scarcity' as rapidly increasing in many marginal environments due to ongoing degradation caused primarily by population growth.

During the last couple of decades, climate change has been added as a key factor in the Malthusian mechanical vicious cycle that is believed to be in operation. This means that the climate-conflict narrative, which has become widely influential, often includes a combination of climate change and population growth as drivers, sometimes also adding ethnic tension to the mix.

Western media and policy-makers have been keen to reproduce this narrative, giving climate change a leading role in explaining conflicts in far-away countries such as in the Sahel. As mentioned in chapter 1, the Sahel was highlighted by the Norwegian Nobel Committee in 2007 as the prime example of such a link between climate change and conflict. The UN Security Council has also discussed this link a number of times over the last 15–20 years, especially with reference to the situation in the Sahel. Several Western countries have brought up this link for discussion and action, while China and Russia have been more sceptical. In addition, some Heads of State in countries with conflicts have been keen to point to climate change as the cause to avoid attention to their own roles in producing the violence (chapter 1).

A major criticism of the privileged attention to natural resource scarcity is that it remains rather vague what the term implies (Gleditsch 1998, Fairhead 2001, Richards 2005). Armed conflicts tend to be about control over land and will therefore necessarily have a resource dimension. It is, however, misleading to thereby conclude that this dimension explains the conflicts. It is also misleading to treat different processes such as environmental degradation, rising population pressure and inequitable access as one overall process of increased scarcity.

Hence, scarcity is treated as a constant variable, while questions of power, politics and history are obscured. Indeed, too simplistic cause-and-effect explanations contribute to diverting attention from more systemic causes that may compromise the interests of the relatively more powerful (Mehta 2010).

Scarcity narratives may in this way serve to justify exclusion of historical land-users. In a study from the Kilombero valley in Tanzania, Bergius et al. (2020) illustrate such a link between the scarcity narrative and its effects on land struggles. Prior to the implementation of a green economy policy (or ecomodernization, see chapter 5) in the valley focussing on large-scale agricultural investments in combination with environmental conservation based on exclusion, pastoralists were portrayed as antagonists to sustainable development, because they were seen as causing widespread overgrazing, while having low economic productivity. This served to legitimise evictions and dispossession, which again caused more land-use conflicts. The further spillover of pastoralists from evictions in Kilombero to other areas resulted in new land-use conflicts, while the same scarcity narrative was reproduced in these areas as well.

This narrative also resonates with widely shared Western and Malthusian perceptions of development problems in Africa. The argument is therefore more a result of Western images of 'the other' than of empirical facts. This 'orientalism' has long Western traditions (Said 1978). It entails regarding smallholder farmers and herders as helpless victims and at the same time pathological abusers of landscapes and natural resources. This kind of thinking helps establish external actors, such as government bodies and international organisations like the UN, in the role of problem solvers (Peluso and Watts 2001).

In contrast to the scarcity narrative and climate determinism, this chapter will present a material political ecology of Fulani–Dogon conflicts in central Mali focussing on historical and political context. As any other violent conflict (Ukraine, Israel/Palestine...), conflicts in the Sahel can only be understood through studying the historical background and the relations of power that are played out by the various actors.

FULANI–DOGON CONFLICTS IN CENTRAL MALI

At 4am on 23 March 2019, an armed group of Dogon traditional hunters[2] attacked and killed 175 Fulani villagers in Ogossagou village in the Seeno plains in central Mali. About half of the casualties in this horrid attack were children. It immediately made international news headlines and was reported as another example of African 'ethnic violence'. But, since Dogon primarily identify as farmers and Fulani[3] as pastoralists,[4] the violence was also presented as resulting from classic tensions between farmers and herders. Media reports often add that these old conflicts are exacerbated by climate change and population growth leading to increased natural resource scarcity.

Ogossagou village in fact consists of two separated sub-villages – one Fulani (Ogossagou Peul) and one Dogon (Ogossagou Dogon). From the 19th century Fulani military power dominated the plains and chased the Dogon in the area back to the nearby escarpment where most of the Dogon population lived. Those who did not want to abandon their farms had to accept to become integrated into Fulani society as Rimaybe – 'slaves' or 'servants' of the Rimbe (Fulani of higher status). After the French colonial power took control over the area early in the 20th century and especially after Mali's independence in 1960, Dogon have gradually moved down again from the nearby escarpment to farm on the plains below (Petit 1998).

On 14 February 2020, Ogossagou was tragically again attacked by Dogon militia (Figure 4.1 and Figure 4.2). This time 31 Fulani villagers were killed. These two assaults are, however, not unique, although the former stands out with its high number of casualties. According to data from ACLED (Armed Conflict Locations and Event Data), 60 per cent of deaths caused by violence in Mali in 2019 were found in this dry savanna belt below the Bandiagara Escarpment and Plateau (called the Seeno, see Figure 4.1) (cited by International Crisis Group 2020).[5]

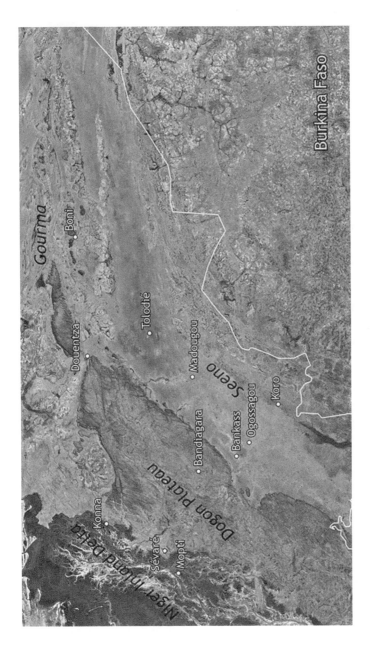

Figure 4.1   Central Mali

Source: Ibrahima Poudiougou.

Figure 4.2   *Hide-out of the Dogon militia Da Na Amassagou in the Bandiagara escarpment in central Mali*

Land Relations and Governance in the Seeno Plains

The Dogon are said to have originated in today's southern Mali in an area south of Bamako and moved during the Mali Empire in the 13th or 14th century to settle in the Bandiagara Escarpment either to escape being converted to Islam by Fulani or because of internal conflicts within their own lineages (Petit 1998, Brandts 2005). The oral history tradition says that they built their villages in the rocky escarpment with a view over the Seeno plains to be safe from slave raids by Fulani, Tuareg and the Mossi and Songhoy kingdoms.

However, in the 15th and 16th centuries, the Dogon were able to move down to the plains to farm, especially in the Madougou area, which is also called Ourokorohi meaning 'old villages'. But later, in the 18th and 19th centuries, they were chased back to the escarpment by a Fulani expansion originating in the delta. Dogon who did not want to leave their farms subsequently became enslaved as Rimaybe (Petit 1998).

In the escarpment and on the plateau above, soils are poor and good farmland is scarce, but with the establishment of French colonial rule in this area from 1905, many Dogon felt safe enough to start the move down again to the plains below to farm (Gallais 1975, Petit 1998). Increasingly, they established cultivation hamlets and opened up new fields (Nijenhuis 2009). In fact, to the Dogon, the plains have historically represented both a 'lost Paradise' and a 'promised land' (Petit 1998).

According to Van Beek (2005: 47):

> At an ever-increasing pace the Dogon swarmed out into the newly opened areas, making fields and founding villages. … in the first decades of the century, the Dogon quickly filled in the empty spots on the map, first along the present border with Burkina Faso with its better soils, and then on the sandy plains closer to the escarpment.

These plains had, up to Dogon colonisation, been used seasonally by Fulani pastoralists who see themselves as the owners of pastures under the authority of the Boni Fulani chiefdom that stretches south from Boni town to also include the sandy plains of the Seeno (De Bruijn and van Dijk 1995). While Dogon since the early 20th century, have transformed vast stretches of pastures to farmland, Dogon farmers would, in some occasions, pay a bundle of millet after the harvest to local Fulani sub-chiefs in recognition of ownership (Nijenhuis 2009). For several decades, the new settlements were, however, not seen as problematic by the Fulani as land was thought to be abundant.

This prompt colonisation has been said to be a result of Dogon 'land hunger' linked to their initial land scarce situation at the escarpment, in addition to population growth (Gallais 1975, Petit 1997). The Dogon population increased from about 100,000 in the early 20th century to at least 300,000 in the early 1970s, according to van Beek (2005). Following the population census of 2009,[6] the Mopti region then had slightly over 2 million inhabitants, of whom 43 per cent were Dogon (and 23 per cent Fulani). This gives a total Dogon population at the time of about 875,000 (and 468,000 Fulani in the region).

Moreover, the replacement of the hoe by the plough in the 1990s facilitated further expansion of farming in the area, which led to a doubling or even tripling of field sizes (Nijenhuis 2009).

The rapid land use change can also be associated with a change in power relations between the Fulani and the Dogon:

> In the nineteenth century, the Fulbe dominated the Dogon, even enslaving many of them. In the course of the twentieth century, however, Fulbe dominance decreased and power differences have since become less clear. In colonial and post-colonial state law, livestock keepers are disadvantaged vis-à-vis farmers since customary rights to land are recognized, whereas pasture rights or territorial rights are not. Sometimes the Dogon (and rimaybe, the former Fulbe slaves) have become more powerful than the Fulbe because of their larger numbers and their relationship with the state apparatus is better (Nijenhuis 2009: 77).

Although the balance of power may have shifted from the Fulani to the Dogon over time, the Fulani are still thought to feel a sense of superiority, 'which seems to keep hampering the Dogon in their dealings with them' (van Beek 2005: 66).

In some cases, Fulani may oppose the expansion of farmland, especially when traditional livestock corridors (*burtol*) are blocked (De Bruijn and Van Dijk 2005). But since farming has usually been supported by the administration of both the colonial and Malian state at the expense of pastoralism, Fulani pastoralists have seldom been able to prevent pastures from being taken over by cultivation (Benjaminsen and Ba 2009).

This situation has resulted in an increasing number of conflicts over land in the Seeno. Such conflicts occur between Fulani and Dogon, but also within each of these groups (De Bruijn and Van Dijk 2005). Conflict resolution may take place with the help of elders or Muslim scholars and usually people have only taken these conflicts to court as a last resort. De Bruijn and Van Dijk (2005: 258) interestingly added – not that long ago – that '(v)iolent conflicts are rare compared to other areas in West Africa'.

As the last few years have shown, this is unfortunately no longer the case. In the following, we try to trace the recent history of these conflicts between Dogon and Fulani in the Seeno. The timeline is based on online sources as well as interviews with resource persons carried out in Bamako in August 2019 together with Malian colleague Boubacar Ba (Benjaminsen and Ba 2021).

After Mali's independence in 1960, an increasing number of Dogon from the escarpment descended to the plains to request access to land from Dogon and Fulani who were already settled in villages. During the colonial period, the French had ruled through the Fulani chief in Boni who had been Chef de Canton over this area including the Seeno (Sangaré 2018). But the independent socialist government of Mali followed both an anti-feudal and anti-pastoral policy, which encouraged farmers to cultivate new areas without asking traditional authorities for permission.

Land conflicts started to emerge in the 1970s and 80s. The first violent conflict took place in Tolodié, which is both the name of a village and a pastoral area. In 1975, a large livestock development programme (Opération de dévelopment de l'élévage dans la region de Mopti, ODEM), with the World Bank as the main funder, had started to dig a total of 11 wells in the Seeno, including one in Tolodié. Pastoral development through improved access to water was the aim of the programme. But by making water permanently available in larger parts of the plains, the water development programme led to a further influx of people to the area. The water rights remained, however, unclear. Discussing the impact of this programme, De Bruijn and van Dijk (1995: 84) found that since 'anyone may use these water sources, herdsmen with livestock and cultivators in search of well-manured land from all over the region settle near these wells'.

While Tolodié had been a pastoral area, the new well also attracted Dogon from the escarpment who settled to establish hamlets and over a few years transformed the area into an agricultural zone. This led Fulani pastoralists to demand the state administration to accept the creation of a *harima* (a pasture reserve for milk cows that do not follow the annual migration). This was agreed, but not respected by farmers who continued to encroach on these pastures. Over time, a conflict emerged that the leader of the pastoral village of Mbana took to court – a case he won in 2002. However, the court decision was not implemented by the administration, due to corruption.

One Dogon even continued to cultivate in the interior of the harima, and one day in August 2002, he was found assassinated. In the night after his body had been discovered, Dogon in the area met, and the next morning they entered Mbana and killed the village chief and seven other people.

After this attack, local Fulani mobilised to evict all Dogon from the Tolodié area, which was an eviction that was accepted by the local administration at the time. But in the following years, Dogon managed to gradually move back to farm in the area.

A few years later, in 2017, some Dogon, without the authorisation of the local Dogon village chief, started to demand fees from Fulani herders for grazing around their hamlets. One Dogon also encroached on the harima with his fields. On 17 June 2017, when visiting the local market with his son, he was killed by some Fulani (others refer to them as 'jihadists').[7]

This led a group of Dogon to organise an act of revenge, which resulted in the burning down of Fulani hamlets and the killing of 40 Fulani in the Tolodié area. The next day, Fulani burnt down Dogon hamlets and killed six Dogon.

This violence in Tolodié in 2002 and 2017 is considered a key event leading to the subsequent escalation of Fulani–Dogon violent conflicts. Another key event was the Dogon attack on Sari on 22 May 2012. Sari is a village situated between Koro and the border to Burkina Faso. In this attack, 350 huts were burnt, 774 cattle taken and 21 Fulani villagers killed in addition to several injured.

At the root of the Sari conflict lies a contestation about a cattle corridor being blocked by farming. This case has also for several years been dealt with by the courts – first the primary court in Koro and then the appeal court in Sévaré – without any resolution, although judgments have been in favour of the Fulani pastoralists being recognised as the first comers in the area. Sari village is said to have been founded by Fulani before the Macina Empire (1818–1862).

The extent of Dogon farming in the area increased in particular after Mali's independence, whereas farmland gradually tended to block the pastoralists' movements and access to pastures.

While the attack in Sari was ongoing, the village chief had called the paramount Fulani chief in Boni for assistance. This request was, however, declined. Boni was at that time occupied by the Tuareg-dominated secular rebel group MNLA, which the chief in Boni was allied with. His refusal to assist Sari was interpreted as a betrayal of common Fulani, not only in Sari. This act further reinforced an existing cleavage between Fulani elites and 'common' Fulani, and it pushed many Fulani pastoralists into the camps of jihadist organisations. The lack of support to Sari when help was called also later motivated jihadist groups in 2016 to attack the Fulani elite in Boni.

The conflict in Sari and its consequences demonstrate the cleavage within Fulani society between traditional elites and more common pastoralists (Benjaminsen and Ba 2019, Jourde et al. 2019). At the same time, it is a representation of a classic farmer–herder conflict over control over space. In addition, this farmer–herder conflict has developed into a confrontation between the Malian army supporting Dogon militia and jihadist groups that many Fulani pastoralists have joined. The Dogon have a strong position in the army with about 25 per cent of the personnel originating from this ethnic group, while Fulani from the northern and central regions of Mali only represent 0.5 per cent of army personnel.

Scarcity and Greening

The greening of the Sahel is also observed in the Seeno plains and on the Dogon Plateau as demonstrated by Brandt et al. (2014) who detected a trend of re-greening with increased tree cover in the area over the period 1982–2010 associated with annual rainfall increases.

While the existence of supply-induced scarcity in the Seeno may be questioned through re-greening, there is still an element of demand-induced scarcity due to population growth. The Dogon's 'land hunger' has historical causes and is the result of a historically produced land scarcity in Dogonland (escarpment and plateau). For centuries, they were cornered in this rocky and land-sparse area because of the danger of being captured as slaves if they descended to the land-abundant plains below.

After this threat disappeared, and encouraged by a policy environment favouring farmers more than pastoralists, Dogon have been colonising the plains and, in many places, transforming a pastoral landscape into farmland. This has led to increasing competition over space, especially where key pastoral land units are encroached upon, such as burtol (livestock corridors) and harima (pastures for milk cows). Hence, pastoralists have often been losing this competition due to an unfavourable policy environment.

Insurgency, Counter-insurgency and Communal Violence

In Mali, militias have often been used in times of crisis to support the armed forces (Diallo 2017). During the Tuareg rebellion in the 1990s, a militia – Ganda Koy – based in the sedentary population of Songhay ethnicity was formed in 1994, which had close links to the military. In addition to self-defence, Ganda

Koy had a clear aim of terrorising civilian Tuareg (Lode 1997, Benjaminsen 2008). Its leadership consisted of former army officers who even wore Malian army uniforms during actions (Boisvert 2015).

Other government-friendly militia in Mali are Ganda Iso, that was established in 2008 to protect Fulani from the Ansongo area against Tuareg attacks, and GATIA (Groupe Auto-défense Touareg, Imghad et Allié) created in 2014 and consisting of Tuareg with a background from the Malian army.

The army has acknowledged that it has provided training and logistical support to these militias (Boisvert 2015). According to Charbonneau (2020), the French Barkhane force also worked with these groups in their anti-terrorism efforts in Mali, which provided the various militias with increased power and further exacerbated inter-community tensions.

Militias are often created by community leaders 'to compensate for insufficient protection from state security forces and thus guarantee protection from violence by rival groups' (Ibrahim and Zapata (2018: 14). Some communities are also allied with jihadist groups to get protection as well as military training and access to weapons, and some communities encourage young men to join these groups (Ibrahim and Zapata 2018).

Da Na Amassagou is the latest addition to pro-government militias in Mali (Figure 4.2). It is usually referred to as a Dogon militia consisting of traditional Dogon hunters as well as former army officers and soldiers of Dogon ethnicity, but also mixed with some combatants who are of Bambara ethnicity.

This militia was established in December 2016 to protect the Dogon population against jihadist attacks (Bourgeot 2019). In 2018, *Da Na Amassagou* experienced a boost in its operational power when it started to receive support from the state after an initiative from the prime minister at the time. The support included automatic hand weapons (e.g. Kalashnikovs) and RPG grenade launchers. This support also allowed *Da Na Amassagou* to set up several training camps in the *cercles* (districts) of Bankass, Koro, Bandiagara, Tominian and Douentza.

The aim of *Da Na Amassagou* also seems to have changed over time from defending Dogon villages to actively attacking Fulani villages. In 2018, the militia reportedly started to systematically attack Fulani villages by setting fire to buildings and chasing the civilian Fulani away. Since Fulani are seen as associated with jihadist groups, this operation may be understood as a state-supported form of counter-insurgency. Several thousand Fulani have, as a result, fled to Burkina Faso to seek refuge.

The strategy of this militia

is not to confront the jihadists who are responsible for the attack but rather to aim at soft targets – in this case, Peul civilians. Also, sometimes armed groups take advantage of the context of conflicts to settle old accounts and disputes, including conflicts over natural resources (Ibrahim and Zapata 2018: 18).

The active state support to Da Na Amassagou continued until April 2019 when public criticism and demonstrations in Bamako led former President Keita to replace the prime minister. A new government approach followed, including ordering the militia in February 2020 to take down five roadblocks in the Seeno.

While the military interventions in Mali, through the UN's MINUSMA mission and the French Barkhane operation in particular, should be understood as expressions of counter-insurgency politics (Charbonneau 2019), the flip side of these endeavours is the government's support to and use of militias for such counter-insurgency. During 2018–19, the active state support to Da Na Amassagou escalated Fulani–Dogon violence and increased communal tensions to a level that seems beyond repair for several decades.

Interestingly, however, a Dogon counter-movement, Da Na Atem, was established in 2020 that condemns the violence perpetrated by Da Na Amassagou and that instead has supported direct negotiations with Fulani and jihadist groups. In these local negotiations without state involvement, jihadist groups have set the condition to expel Da Na Amassagou, to ban arms locally, to introduce sharia-based family laws and taxes, and a ban on any contact with the Malian state and army (Ba and Cold-Ravnkilde 2021). While the foreign military intervention has been allied with the Malian army responsible for serious human rights crimes, which again has supported violent militias such as Da Na Amassagou, the emergence of these locally initiated peace deals further demonstrates the complexities of actors and interests in conflicts in Mali, as opposed to simplified media and policy narratives.

CONCLUSIONS

Farmer–herder conflicts in the Sahel tend to be presented by external observers as either driven by ethnicity or by increasing resource scarcity caused by climate change or population growth – or a combination of these factors. While these elements may be relevant, which we also see in the case of Fulani–Dogon violence in Mali, narratives about ethnicity or scarcity also serve to simplify explanations of conflicts and to gloss over more complex root causes linked to historical and contemporary land relations.

In order to more fully understand such land struggles that escalate into violence, it is therefore necessary to study the material politics of land governance in a historical context. This will lead to explanations that move beyond simple causal chains where the Sahel is seen as an arena with decreasing resource scarcity affecting livelihoods that again trigger migrations or conflicts.

In the case of the Fulani–Dogon killings in the Seeno plains in central Mali, which has been going on for the last few years, a series of longer historical and more recent facts and processes have led to the current situation. First, in periods, the Dogon have had to seek refuge in the land-scarce Bandiagara escarpment and on the Dogon Plateau in order to avoid being captured as slaves if they descended to the land-abundant savanna below. While the Dogon managed to settle on the plains to farm in the 15th and 16th centuries, they were later chased back to the escarpment by a Fulani expansion in the 18th and 19th centuries.

Dogon increasingly moved back again to the Seeno with French colonisation, and especially after independence, when Mali's new socialist government followed an anti-feudal and anti-pastoral agenda that motivated and increased the speed of the Dogon migration to the plains. In addition, Mali's land policies and laws have continued to favour farming at the expense of pastoralism. This has encouraged the transformation of a pastoral landscape into farmland including the blocking of livestock corridors and encroachment on key pastures, which have led to the emergence of conflicts between farmers and pastoralists. In addition, access to pastoral land has often been lost through corrupt behaviour by government officials and traditional elites, which has increasingly left herders with a feeling of disempowerment and marginalisation.

After the emergence of a national crisis in Mali from 2012 and the jihadist take-over of the northern and central parts of the country, many Fulani pastoralists have been attracted to an anti-government, anti-elite and pro-pastoral jihadist discourse. While Dogon had been increasing their power relative to the Fulani since independence until 2012, this power balance began to shift back in favour of the Fulani with the arrival of jihadist groups in the area.

The creation of the Dogon militia, Da Na Amassagou, that has attacked and killed a large number of Fulani villagers, can be seen as a response to this threat of a move back towards Fulani power that many Dogon fear would again leave them with a position as subordinates. In addition, since the Malian army has failed to defeat the jihadists, the state has supported the militia with heavy hand weapons and training to fight the insurgent groups in its place. Hence, the recent escalation of this farmer–herder conflict can be seen as a manifestation of the jihadist insurgency and its counter-insurgency.

Finally, while ethnicity and scarcity are clearly relevant in this case, these factors hardly explain the conflicts. We also see that there are conflicts among the Fulani – between elites and subordinates and between Fulani pastoralists adhering to different jihadist groups who fight over grazing rights. There is also a cleavage among the Dogon between a violent militia and a new organisation seeking negotiations with Fulani and 'jihadist' groups.

While the Dogon was historically trapped in a land scarce situation, this changed especially after independence. More recently, increased land-use has filled up space in the Seeno creating land scarcity, but this seems more to be a scarcity created politically through failed land governance by the state rather than being absolute physical scarcity. The Sahel, including the Seeno, has also been re-greening during the last few decades in contrast to popular media and policy presentations.

In addition to demonstrating the complex background to the current crisis in Mali and how simplistic narratives about its causes are not helpful for our understanding of this crisis, the case also shows how views of the enemy as 'terrorists' or 'jihadists' are dangerous and able to further fuel violent conflicts.

NOTES

1. The term 'conflict' has been used in many different ways. It may refer to tension between groups of resource users, between individuals, or between resource users and the state. It has also been used on both oral disagreements over land-use taken to court, evictions of resource users, theft, raiding of livestock and large-scale violence between groups involving multiple killings (Hussein et al. 1999).
2. The traditional role of hunters in Dogon society is to defend the village in general and to police the village territory outside the cultivated areas. This includes controlling wood cutting and wildlife hunting.
3. The Fulani are also referred to as Fulbe or in French – *Peul*.
4. Although Dogon mostly identify as farmers and many Fulani as pastoralists, most rural people in central Mali have diversified their livelihood strategies to reduce vulnerability. This means that many Dogon keep livestock and many Fulani grow millet in particular (Nijenhuis 2009).
5. ACLED data indicate that totally more than 2000 people were killed in this wave of violence only in 2019 and the first four months of 2020 (International Crisis Group 2020).
6. Recensement général de la population et de l'habitat du Mali (RGPH): résultats définitifs, Tome 1: série démographique, Institut national de la statistique, novembre 2011.
7. There are different versions of this story. Some say that this Dogon had killed two herders from the Boni area who accused him of stealing livestock. Others say that he was an important hunter who was already organising a Dogon militia to prepare attacks on Fulani.

5. Green transformation, afforestation, land dispossession and context

ECOMODERNISM, DEGROWTH AND AFFORESTATION

In the last decade or two, frequent climate-related events (hurricanes, wildfires, floods, droughts, heat waves, ice melting) as well as new climate research and recent IPCC reports with steadily more urgent messages have led to an increasing global realisation that the climate crisis represents an emergency that necessitates swift action.

However, ideas about what this action should consist of differ widely. At the Rio + 20 conference in Rio de Janeiro in 2012, 'green economy' was launched as a new instrument to bring about such action (UNEP 2011, OECD 2012, World Bank 2012). Supported by powerful economic and political interests, a green economy discourse has been further bolstered since the Rio meeting. It tends to have a focus on market-based and technological solutions to foster triple-win scenarios (climate mitigation, biodiversity conservation, livelihoods development) with the help of green economic growth.

This dominant approach to green transformation today informs policies in most countries in the Global North and often also in the South. The approach has been further articulated in 'An Ecomodernist Manifesto' published by The Breakthrough Institute in Washington DC (Asafu-Adjaye et al. 2015). The manifesto outlines key steps humanity must take to 'allow for a good, or even great, Anthropocene' (Asafu-Adjaye et al. 2015: 6). The main aim is to decouple economies from environmental impacts through intensified agricultural practices based on large-scale high-tech farming combined with increased urbanisation and conservation of 'unused' rural land. In contrast, modernisation is not seen as possible in a subsistence-based agrarian economy. The implication is that countries in the Global South should urbanise and replace small-scale farming and pastoralism with large technology-intensive farms that can feed urban populations (Bergius et al. 2018).

This means that ecomodernism implies a green growth approach focused on the development of green cities combined with the conservation of wilderness, dispossession of small-scale farmers and pastoralists that are seen as inefficient producers, and the development of modernised large-scale agribusiness.

81

Most people should, in this thinking, live in cities, while rural areas are used for large-scale farming and the conservation of 'unused' areas. In other words, ecomodernists do not seem overly concerned about the local social and economic consequences of their technology and market optimism.

Ecomodernism also seems to be a leading premise in most IPCC scenarios with reliance on green economic growth, large-scale depopulation of rural areas, accelerated urbanisation, reductions in cropland and pasture and afforestation[1] of enormous areas (Bluwstein and Cavanagh 2023). While the IPCC has also recently acknowledged degrowth, defined as a down-scaling of production and consumption in the Global North, as potentially an alternative approach to societal transformation (Ara Begum et al. 2022), ecomodernist ideas still dominate within IPCC reports.

In broader public and academic environmental and climate debates, there is, however, clearly a growing interest in the more radical degrowth ideas, although these ideas still have marginal political influence, and tend to be confined to activist and some academic environments.

According to leading degrowth advocate Jason Hickel,[2] degrowth implies planned reduction of aggregate resource and energy use in high-income countries to stay within planetary boundaries. Such reduction is necessary as there is currently an overshoot of these boundaries. Available evidence shows that GDP growth is tightly coupled with resource use, ecological damage and biodiversity loss. The current overshoot of planetary boundaries is primarily driven by rich countries that are responsible for 92 per cent of the excess emissions with a material footprint four times the sustainable level, according to Hickel. Green growth in the North based on renewable energy such as solar and wind power will also mean increased extractivism in the Global South, which supply most of the materials for these technologies.

This means, Hickel argues, that the necessity for rich countries to degrow their economies is about more than just ecology and climate change. It also implies a call to decolonise global resource use patterns and to end the Global North's appropriation of resources in the Global South. Much of the current growth in the North is fed by the appropriation of resources in poor countries that props up consumerism and elite capture. The implication is that the North should stay within planetary boundaries, while the South should be able to meet human needs.

Finally, there is no evidence of long-term absolute decoupling, according to Hickel. Such decoupling is the basic premise of green growth and ecomodernism. This is also unlikely to happen in the future due to the Jevons Paradox (the rebound effect),[3] he argues.

As mentioned, most IPCC scenarios are built on the premise of continued economic growth in industrialised countries. To mitigate this growth, the scenarios rely heavily on bioenergy with carbon capture and storage (BECCS)[4] and afforestation. But these methods to remove atmospheric carbon dioxide will, in order to offset planned growth, need access to enormous land areas, up to three times the size of India, which will mostly be appropriated from the Global South (Hickel 2021). For these reasons, scaling down resource use to sustainable levels in rich countries is necessary.

However, as Cavanagh (2021) points out, Hickel has also curiously written positively about the Low Energy Demand (LED) scenario of the IPCC (Grubler et al. 2018) as an alternative scenario with the aim of staying within the 1.5-degree climate target. Hickel sees this as a degrowth scenario, since it combines a 40 per cent reduction in global energy use by 2050 with an avoidance of BECCS. The problem is, however, that instead of relying on BECCS, the LED scenario depends on large-scale afforestation at the scale of 646 million hectares by 2100. This is equivalent to about twice the size of the Indian landmass. Most of this afforestation is thought to take place in Africa where there is believed to be large spans of available land. Such mega-scale conversion of productive farmland and pastures to mitigate climate change carries a high risk of increased conflict following both displacement of populations and potential resistance from dispossessed people (Cavanagh and Benjaminsen 2022).

Rather than being a critique of degrowth, this example illustrates the scale of the climate problem, the need for radical reductions in the use of energy and materials, the potential threat to climate justice of some commonly trusted carbon dioxide removal technologies, and the danger of relying on these methods (McElwee 2023).

In a similar vein, Deprez et al. (2024: 484) report that:

> The latest IPCC Working Group III (WGIII) report estimates the upper 'technical mitigation potential' of BECCS and A/R[5] at 11.3 and 10 gigatonnes of CO_2 per year (GtCO$_2$/year), respectively. Together, this could require converting up to 29 million km^2 of land – over three times the area of the United States – to bioenergy crops or trees, and potentially push over 300 million people into food insecurity.

Despite the fact that large-scale afforestation is an extremely land-hungry form of mitigation with a large potential to negatively affect livelihoods and increase land-use conflicts, afforestation is still widely seen as a simple and efficient way to fight climate change. For instance, an influential article in Science that used satellite imagery and machine learning, and received wide media coverage, stated that:

> The restoration of trees remains among the most effective strategies for climate change mitigation. We mapped the global potential tree coverage to show that 4.4 billion hectares of canopy cover could exist under the current climate. Excluding existing trees and agricultural and urban areas, we found that there is room for an

extra 0.9 billion hectares of canopy cover, which could store 205 gigatonnes of carbon in areas that would naturally support woodlands and forests. This highlights global tree restoration as one of the most effective carbon drawdown solutions to date (Bastin et al. 2019).

The article, however, also immediately received several critical responses. One of the responses pointed out that 205 gigatonnes is equivalent to 20 times current annual fossil fuel emissions and about one-third of total historical anthropogenic emissions (Veldman et al. 2019). Such numbers are grossly exaggerated, these critics argued. For instance, Bastin et al. (2019) assumed that treeless areas have no soil organic carbon, while humid tropical savannas store 86 per cent of all carbon in soils, deserts and shrublands contain 5.1 tC/ha in soils, and even in boreal forests soils contain 64 per cent of the carbon in the ecosystem. As planted trees may outcompete grass, tree planting in savanna areas is often not very helpful in terms of carbon sequestration.

Another critical point not taken into account by Bastin et al. (2019) is that forest plantations, especially of eucalyptus and pines, are highly vulnerable to fires and may even become more vulnerable with higher temperatures, while the soil carbon in grasslands and savannas is persistent to fire (Bond et al. 2019).

Following all the scientific critique, Bastin et al. (2019) later corrected their article in Science with an erratum in the journal. The media attention had, however, already made a huge impact on public opinion. The Guardian had, for instance reported (4 July 2019) that 'Tree planting "has a mind-blowing potential" to tackle the climate crisis'.

The newspaper followed up by citing the lead researcher of the study, Professor Tom Crowther at the Swiss university ETH Zürich, who stated that 'this new quantitative evaluation shows [forest] restoration isn't just one of our climate change solutions, it is overwhelmingly the top one'.

Commenting on this study, former UN climate chief Christiana Figueres said: 'Finally we have an authoritative assessment of how much land we can and should cover with trees without impinging on food production or living areas. This is a hugely important blueprint for governments and the private sector.' Also supporting such large-scale afforestation, René Castro, assistant-director general at the UN Food and Agriculture Organisation, said: 'We now have definitive evidence of the potential land area for re-growing forests, where they could exist and how much carbon they could store.'

Other headlines said 'Massive forest restoration could greatly slow global warming' (Scientific American 4 July 2019), 'How to fight climate change? Plant a trillion trees' (Voice of America 14 July 2019), 'Best way to fight climate change? Plant a trillion trees' (Al Jazeera 4 July 2019), and 'How to erase 100 years of carbon emissions? Plant trees – lots of them' (National Geographic 4 July 2019).

Several global tree planting campaigns have followed up on these ideas. The Bonn Challenge, initiated by the German government and the International Union for Conservation of Nature (IUCN), stated that it had 'a global goal to bring 150 million hectares of degraded and deforested landscapes into restoration by 2020 and 350 million hectares by 2030', mainly through the means of tree planting.

The African Forest Restoration Initiative (AFR100) is a tree planting programme affiliated with The Bonn Challenge. It aims 'to bring 100 million hectares of land in Africa into restoration by 2030'. So far 34 African countries have joined the campaign according to the AFR100 website. One of the countries is Mali where there is only one project. The project website states that:

> The Sahel Region has lost millions of hectares of easily accessible farming land to the desert, thus creating food insecurity and loss of income for thousands, some due to climate change and others due to continued pastoral activities. 'Herders who bring animals and cut trees bring conflict among farmers and this is a major problem in our areas', (the project leader says). (The project will first aim) to bring together farmers and herders and then to reforest most of the parts that were deforested as a result of the cattle ... (and it) intends to scale up regreening practices in San and Tominian regions in Mali, build their capacity to scale up and work with more farmers to produce and grow more trees over time.

The Nature Conservancy's Plant a Billion Trees campaign is another major afforestation and reforestation programme. Their tree planting is glossed as part of 'natural climate solutions'. Referring to the World Economic Forum's global initiative to grow, restore and conserve one trillion trees around the world and, anticipating critiques from conservationists, The Nature Conservancy states that the fact that such a 'campaign to plant a trillion trees (takes) centre-stage in Davos is a sign that the wind is in our sails. The fact that political and business leaders are also jumping on board does not make the science around natural climate solutions any less true'.

However, political leaders are attracted to large-scale tree planting because of the relative low costs and apparent quick fixes involved (Deprez et al. 2024). In addition, using trees and forests in faraway countries as a key component in climate mitigation by rich countries is seen as a measure that takes attention away from controversial discussions at home about consumption levels that tend to cause dissatisfaction among domestic electorates (Benjaminsen and Hiernaux 2019).

Governments in fossil fuel-producing countries may, in particular, be drawn towards forest and tree-based solutions to climate change. Norway is, for instance, by far the biggest donor country of initiatives to use tropical forest conservation as a climate solution through so-called REDD (Reduced Emissions from Deforestation and forest Degradation). While it is important to arrest deforestation, REDD has also led to dispossession of small-scale forest users without compensation (Svarstad and Benjaminsen 2017, Asiyanbi and Lund 2020, Benjaminsen and Svarstad 2021). These social costs are glossed over by Norway at, for instance, climate meetings where such action is promoted.

In a similar vein, the royal family in the United Arab Emirates created the company Blue Carbon just months before the climate summit COP 28 in Dubai in 2023. Already before the climate meeting, the company had managed to establish agreements with five African countries – Zimbabwe, Zambia, Kenya, Liberia and Tanzania. These agreements involve forest conservation and afforestation on land equivalent to the size of the United Kingdom. In this way, fossil fuel-producing countries, such as Norway and the Emirates, hope to use trees to greenwash their oil and gas production through the sales of carbon credits.

Also African governments are clearly attracted to such projects, because of the initial financial investments including promises of job creation and the funding of infrastructure (Bond et al. 2019).

As an example of the political attraction in rich countries to tree planting, The Washington Post reports (2 August 2023) that 'Republicans want to plant a trillion trees':

> 'God has blessed America with resources', (House Speaker Kevin) McCarthy said. 'If we have the ability to produce those resources, America will be stronger and the world will be safer.' Smoke from Canadian wildfires hung in the air as McCarthy spoke. Asked about his plans to prevent further fires and other disasters fueled by climate change, the speaker suggested a strategy popular among Republicans: Plant a trillion trees. ... The plan has some prominent backers. President Donald Trump announced in 2020 that the United States would join a global initiative to plant a trillion trees, despite his antagonism toward climate science. The chairman of the House Natural Resources Committee has introduced legislation to plant a trillion trees as 'a comprehensive, practical solution to the climate issues we're facing today'.

The risk of this strategy is, as pointed out by Deprez et al. (2024), that the consequences of climate change will be even more severe in the future:

> Many governments and industries are relying on future large-scale, land-based carbon dioxide (CO_2) removal (CDR) to avoid making necessary steep greenhouse gas (GHG) emission cuts today. Not only does this risk locking us into a high overshoot above 1.5°C, but it will also increase biodiversity loss ... Such CDR deployments also pose major economic, technological, and social feasibility challenges; threaten food security and human rights; and risk overstepping multiple planetary boundaries, with potentially irreversible consequences.

In addition, what is often not realised or reflected in debates about using trees and forests for climate mitigation is that there are two types of carbon cycles. Here is what 41 scientists write in a blog article on '10 myths about net zero targets and carbon offsetting':

> Fossil fuels are part of the slow carbon cycle … Nature-based solutions are part of the fast, biological carbon cycle, meaning that carbon storage is not permanent. For example, carbon stored in trees can be released again by forest fires. Fossil emissions happen today, while their uptake in trees and soils takes much longer … The carbon cycle has two parts: one fast cycle whereby carbon circulates between the atmosphere, land and seas, and one slow cycle whereby carbon circulates between the atmosphere and the rocks which make up Earth's interior … Fossil fuels (coal, oil and gas) come from rocks (part of the slow cycle) (Skelton et al. 2020).

This means that trees are part of the fast carbon cycle as they may easily burn releasing the carbon back into the air. Fossil fuels, on the other hand, come from the deep underground and were formed as organic deposits millions of years ago. Burning these fuels obviously increases the amount of carbon in the atmosphere, and compensating this process with growing trees remains extremely uncertain. In addition, political change may also represent a risk by introducing regimes that are less favourable to using trees and forests for climate mitigation, which again may lead to forest loss.

Despite these obvious problems with tree planting as climate mitigation, it remains exceedingly popular among Northern governments and industries. Such greenwashing, which has questionable effects on the climate, and which causes serious social and economic problems in many localities, Africa in particular, tends to gloss over the real problems of fossil fuel consumption as a cause of global warming.

These projects often target savannas and grasslands with low tree cover, 'often erroneously assuming that these areas are deforested and degraded and therefore represent opportunities for restoration' (Parr et al. 2024: 701). In addition, to wrongly assuming that the savannas are degraded (see chapter 2), they also assume that they are unused and without any ownership attached to them. This sums up to afforestation being a neo-colonial climate solution (Briske et al. 2024). To understand the context of afforestation initiatives and their impacts on land-use, it is therefore important to take a historical glance at issues of land rights and custom.

COLONIALISM, LAND RIGHTS AND CUSTOM

In the colonial period in French Sudan, the State and individuals with title deeds were the only possible landowners. This idea came from the French Code Civil or Code Napoléon from 1804, stating in article 539 that 'Tous les biens vacants et sans maître ... appartiennent au domaine public' ('All vacant goods without a master belong to the public domain') and in article 713 that 'Les biens qui n'ont pas de maîtres appartiennent à l'Etat' ('Goods without a master belong to the State') (see Rochegude (1977) for a review of colonial law in French Sudan). This policy implied that only the State and individuals with title deeds could hold formal property rights to land. However, over time, customary rights were given some recognition within the colonial system, but still custom was only acknowledged as giving the status of use rights.

As Sûret-Canale (1962: 255–56) points out:

> This application of the Roman property concept did not arise, as is sometimes alleged, from a lack of knowledge of the African situation. All colonial authors recognised that in tropical Africa 'not an inch of land is without an owner'. This was clearly stated in 1920. In a report adopted unanimously by the general council of the Ivory Coast, a councilor, J.B. Mackey, recalled: Not a single square metre of land can be considered as ownerless. Vacant, no doubt, but most often momentarily, periodically, through the crop rotation system, by the intermittent use of pastures, etc. But ownerless? Definitely not. No land in Africa can be so considered. The tribes agreed upon borders among themselves, borders which generally followed natural features such as water courses and mountain ranges; or borders were fictitious ones, their position determined with the aid of landmarks usually placed along routes or taking the form of rocky outcrops. If tribes had borders, a fortiori sub-tribes or cantons and smaller territorial units also had precisely defined borders. This applies even on the scale of the village and the family.

In other words, it was not lack of knowledge that erased African customary land rights, in contrast perhaps to contemporary afforestation projects (Turner et al. 2021), but custom was rather seen as an inconvenience and an obstacle to the extraction of resources that was at the heart of the colonial enterprise. In French Sudan, cotton was key to the French colonial interest.

As we shall see, the French colonial government used two very different approaches to provide cotton from French Sudan to the textile mills in cities like Lille, Rouen, Marseille and Nîmes. One approach was based on smallholder production in villages in southern Mali and the other consisted of a large land grabbing project to produce cotton in the area that became known as 'Office du Niger'. The former approach, that largely respected customary land rights, was eventually more successful in terms of both providing an income for local farmers and increasing cotton production, although this process only took off after independence. The latter scheme led to forced labour as well as dispossession of villagers and pastoralists under colonial governance, and more recently to further land dispossession and elite capture.

At the end of the 19th century, almost all the cotton processed by the textile industry in France was grown in the USA. However, the Americans were developing their own textile industry, which led the French to fear losing their chief source of the raw material. The French industry, therefore, looked for new areas that could supply the cotton it needed, and the West African Sahel was picked out as a promising new area for cotton supply.

There were several reasons for this choice: the local farmers, representing a cheap labour force, were already cultivating cotton; the climate was favourable for cotton production; and transport would be possible through the use of the Senegal and Niger rivers and the Dakar-Bamako railroad under construction (it was finished in 1904).

In order to encourage farmers to grow more cotton, this crop was accepted as tax payment from 1896 onwards (Sanogo 1990). As early as 1893, foreign varieties of cotton had been tested in the Kayes area. In 1903, the 'Association Cotonnière Coloniale' (ACC) was created by the French textile industry. The same year, a representative of the ACC arrived in San and distributed a large amount of long fibre American cotton seeds to villagers (Benjaminsen 2001).

However, it soon proved difficult for the ACC to receive stable supplies from local farmers. The main reason was the existence of a local cotton market with higher prices than the French were willing to pay (Fok 1994, Roberts 1996).

Disappointed by the modest and uncertain supplies, colonial authorities made cotton cultivation compulsory in 1912, and cotton became 'la culture du Commandant'. Each village was obliged to supply a certain amount (10 kg per taxable person). This level, however, was not attained because of local passive resistance (Bassett 1995 and 2001, Roberts 1996), due to the fact that forced cotton cultivation competed for manpower with the production of food crops. Peasants therefore tried to reserve the strongest hands for food crops (Rondeau 1980), in addition to using the 'weapons of the weak' (Scott 1985) through general foot-dragging and neglect of cotton fields (Bassett 1995 and 2001).

Despite efforts by the colonial government to distribute new varieties, introduce ox-ploughs and make cotton production compulsory, the production of cotton in French Sudan remained disappointing seen from the colonial point of view, and exports of cotton from the colony to France fell during the 1930s and 1940s.

The turning point in the region's cotton production came with the increase in world cotton prices due to a bad harvest in the USA in 1950. In addition, from 1952 cotton production was based on a guaranteed price announced in advance by the colonial cotton company (CFDT). The result was an increase in commercialised cotton in French Sudan from only 150 tons in 1952 to 3900 tons in 1958. This was only the beginning of a rapid rise in the production of export cotton. In 1972, it reached 68,000 tons and in 2003 593,000 tons (Benjaminsen et al. 2010).

In 2023, Mali regained its earlier position as Sub-Sahara Africa's leading cotton producer with a production of 690,000 tons. This rapid rise in production since the 1950s has been labelled a 'success story' with reference to

Malian cotton as 'white gold' – a success that has been largely attributed to the CFDT/CMDT model (Benjaminsen 2001). After independence, the colonial cotton company (Compagnie Francaise pour le Développement des Textiles) had become Compagnie Malienne pour le Développement des Textiles, 60 per cent of which was owned by the Malian state and 40 per cent by French capital. The relative success of this model has been achieved because farmers have received information about the farmgate price from a guaranteed buyer before the agricultural season, in addition to receiving, in some periods, subsidised supplies of fertilisers as well as other forms of extension support. In addition, and perhaps most importantly, farmers have been able to produce cotton on their village land without the land dispossession that is frequently associated with most large-scale agricultural schemes.

The 'Office du Niger', on the other hand, stands in stark contrast to this development model. The planning of this irrigation scheme started in the 1920s, and in 1932 it was officially opened with its headquarters in Ségou. Most of the original scheme was located in what is called 'le delta mort' in the southern part of the inland delta of the Niger river, which consists of drylands adjacent to the regularly flooded delta. By 1945, the project had forcibly resettled 30,000 villagers (Filipovich 2001). In addition, pastoralists had lost access to pastures, and old livestock corridors had been blocked creating numerous conflicts with the project (Touré 2022).

The initial aim was to irrigate 1,850,000 hectares and protect the land from an encroaching Sahara, but only 60,000 hectares were finally realised by the colonial authorities. Since cotton is a labour-intensive crop and the project area had a low population density, forced labour was introduced early in the project period. The resettled farmers and workers, however, resisted in various ways – through growing food rather than cotton and by failing to follow other project directives (Van Beusekom 1999). In addition, guards prevented farmers from leaving the scheme. This added up to the project being a 'monumental failure' (Filipovich 2001).

More recently, in the 2000s former President Amadou Toumani Touré (ATT) aimed to revive this project and attract foreign capital to expand the area under cultivation. This led to the arrival of numerous foreign investors from a variety of countries, the construction of new irrigation canals and the dispossession of many villagers and pastoralists, including the bulldozing of village settlements (Oakland Institute 2011, Larder 2015, Toulmin 2020).

One such case of land investment is the Malibya project, which emerged in 2008 as a deal between the former Libyan Leader Muammar Gadhafi and ATT. Through the Libya Sovereign Wealth Fund, the Libyan state was granted access to 100,000 hectares. The project included the construction of a 40km long canal, one of the largest in Africa, and production of hybrid rice (Oakland Institute 2011, Larder 2015).

Malibya was also granted very generous conditions including exemption from fees to use the land for 50 years, negligible fees for water extraction and exemption from company tax for eight years. However, in the construction of the canal, houses in villages were razed, some gardens and orchards were bulldozed, livestock corridors were blocked, and a cemetery was dug up exposing human remains. By July 2009, out of 150 households affected by the initial construction, only 58 had received any compensation (Oakland Institute 2011, Larder 2015). Since the crisis in Libya in 2011, and the crisis in Mali from 2012, the project has, however, been on hold.

While such large land grabbing projects erase 'customary rights', there is also a more nuanced discussion about what 'custom' implies. The Dina code and its associated rights are today seen as part of 'customary law' in Mali in contrast to Roman law introduced through the French colonial system. This was, however, a system of land rights formalised by the Macina Empire in the 19th century (Benjaminsen and Lund 2002).

In this discussion about the relationship between state law and customary law, one currently influential position holds that conceptions of 'tradition' and 'custom' are largely results of constructions or inventions by colonial authorities (Ranger 1983). This literature has provided important insights into the practice of indirect rule, colonial co-option of African chiefs and the exploitation of the peasantry by the combined force of the colonial state and local elites. However, these views are essentially based on empirical work in southern Africa. In studies from other parts of the continent, questions have been raised about the validity of these conclusions. Informed by his own historical studies from Tanzania, Spear (2003: 3) argues that 'the case for colonial invention has often overstated colonial power and ability to manipulate African institutions to establish hegemony'. It therefore makes little sense to talk about 'invention', because '(t)radition was both more flexible and less subject to outside control than scholars have thought' (p. 26). While Spear focused on chieftaincy and ethnicity, Lentz (2006) has extended the argument to the issue of land tenure based on studies in West Africa. She contends that not all statements about past authority in land matters are results of recent 'inventions'.

In Mali after independence, the Land Code of 1986 (Code domanial et foncier) only acknowledged property rights in the case of individually held title deeds. Customary rights were deemed as use rights with a much weaker status than titled land. This Land Code was later replaced by a Land Ordinance in 2000 (Ordonnance du 22 Mars 2000 portant code domanial et foncier) and an Agricultural Land Law of 2017 (Loi portant sur le foncier agricole). The Land Ordinance allows for the titling of land also in the name of groups (*collectivités*), either in the form of residential lineage groups, villages, nomadic groups, or communes. However, the definition of a group often remains a problem, raising issues of permissible entry and exit as generations succeed each other, and is difficult to solve once and for all. As long as the groups have not registered their land, they can only hold customary use rights.

The Agricultural Land Law states that no land held under customary law shall be included under state land. The law also provides for the possibility to register customary land rights by creating two new land titles – 'customary land certificates' and 'certificates of land possession', and the law also recognises the rights for rural communities to collectively own land.

As the application of this law is still limited, due to the unstable political and security situation, it is uncertain what the law's impact on pastoralism will be as pastoralists depend on flexible arrangements. Exclusive property rights including for groups may often represent a problem to pastoral use, and pastoralists are themselves often not interested in receiving exclusive rights to land. They only need access at crucial periods of the year (Benjaminsen 1997, Berge 2001, Scoones 2021).

This need of access to seasonal pastures clashes with not only large-scale irrigation projects. The widespread plans for afforestation on the African continent are also bound to create dispossession and conflict. We have already seen afforestation with these consequences both for pastoralists and small-scale farmers. This chapter continues with two such examples – the first from East Africa and the other from the Sahel.

GREEN RESOURCES IN EAST AFRICA

The investments of the Norwegian forest company Green Resources represent perhaps the longest experience in time of afforestation in Africa in recent decades. The impact of these investments may therefore provide lessons for forthcoming planned afforestation on the continent and in the Sahel.

Green Resources (GR) was established in 1997 and has acquired large plantations[6] in Mozambique, Tanzania and Uganda. The company claims to be Africa's largest forest company with 45,000 hectares of standing forest, 3,500 employees and 80 shareholders (mostly Norwegian) that by 2017 had invested about 300 million USD. In addition, the company has received Norwegian development aid through Norad and a loan of 17.9 million USD from Norfund (Bergius et al. 2018).

The company says it plants ten new trees for each tree harvested, and that plantations are only established on 'low value grassland or degraded forest-land'. Moreover, the company presents its activities within a sustainable development framework that implies a focus on community development and local benefits. This includes village afforestation, the construction of school buildings, roads and village halls and offices, in addition to providing local employment. According to the former CEO 'no carbon mitigation activity creates larger economic benefits for the rural poor than afforestation'.

Situated in the Southern Highlands of Tanzania are three of GR's oldest and biggest plantations that were established in 1997. The forest plantation covers around 74,000 ha. Part of the plantations produces timber and the rest are planted to sequester carbon and generate carbon credits. These plantations have been registered under the voluntary carbon standard (VCS) selling credits on the voluntary market.[7]

In total, land has been allocated from six different villages, two villages for each plantation. The villages have had little to say in this process, as most of the land was negotiated before the Village Land Act of 1999, which means that at the time the government and not the village council managed the land. This has led to some of the villages losing more than 33 per cent of their land, which was set as the limit in the Village Land Act (URT 1999) for how much a village can give away to an investor. For example, one of the villages, Uchindile, lost almost 60 per cent of its land to Green Resources (Refseth 2010).

The company has received leasehold for the land for 99 years from the government. According to the Village Land Act, villages cannot lease land directly to investors. The land needs first to be converted to general land, which is managed by the government. This means that when the period for leasehold is out the land is returned to the government and not to the village. In return for giving away land villages are promised employment, development of infrastructure and support to community projects. In addition, GR promised 10 per cent of the total revenue from selling carbon credits to the villages (Refseth 2010).

Recent and earlier reports have shown that the benefits from GR projects to the local communities have not been fulfilled as promised (e.g. Refseth 2010, Point Carbon and Perspectives 2008, Karumbidza and Menne 2011). Refseth (2010) found, for instance, that of the promises made in 1997, about one third had been honoured in 2009.

The only roads that had been constructed were roads in the plantation itself, which were not directly benefiting the villages. Despite promises of access to safe water, no efforts to supply water to the villages had been made. Lastly, support to community projects had been slow and barely existed (Refseth 2010). In sum, one main problem with the approach used by GR is that local benefits are not transparently stated in written contracts. Hence, benefits do not become rights that communities hold, but are merely subject to charity from the company (Bergius et al. 2018).

This has also led to considerable resistance within the project area consisting, for instance, of widespread thefts of various tools and small equipment as well as the setting of fire on parts of the forest. This has happened several times and does not seem to be accidental.

The experiences with the GR plantations in Uganda and Mozambique are not any better. Lyons and Westoby (2014) argue that the GR project in Uganda is an example of 'carbon colonialism' where GR obtained control over two forest reserves (of 4500 and 2669 hectares). The plantations in one of these reserves were approved by the Clean Development Mechanism (CDM) and carbon credits were sold to the Swedish Energy Agency (SEA) with a contract for the period 2012 to 2032 with a value of about 4 million USD (Hajdu et al. 2016).

CDM rules require that it should be demonstrated that 'lands to be afforested or reforested are degraded and (that) the lands are still degrading or remain in a low carbon steady state'. It is up to the organisation responsible for the afforestation to provide evidence of degradation. This means that GR provided the evidence that the forest reserve in question in Uganda – the Kachung forest – consisted of degraded savanna and that this was accepted by the CDM. This degradation claim was a key motivation for the project, which was apparent in the GR documentation, in the CDM certification and in the SEA's promotion of the project in Sweden (Hajdu et al. 2016).

'Degradation' was defined in these project presentations simply as the loss of trees – a definition that was questioned by Hajdu et al. (2016). In addition, if the purpose is to store carbon, a grass savanna has, as previously mentioned, most of its carbon stored underground, and planting trees in such areas may lead to a loss in the underground carbon. Whether the planted trees will compensate for this loss is also uncertain, as trees tend to burn, while carbon below ground is protected against fire.

As demonstrated by Hajdu et al. (2016), the environmental history of this forest reserve is also more complicated than what GR and SEA pretended. Before the 1930s, when the first plantations in the area took place, the Kachung 'forest' was a savanna used for communal grazing. In 1952, the area was gazetted as a colonial forest reserve earmarked for further afforestation. However, the plans for tree planting were only partially followed and after independence, in the 1960s and 70s, the Ugandan government encouraged people to convert forest reserves to farmland. In Kachung, this process started in 1966 and continued into the 1970s. The area had in fact fertile soil and attracted many farmers. Nevertheless, formally Kachung was still a forest reserve under the state and cultivating the land gave no formal rights. This fact created conflicts when the Ugandan government by law in 2003 invited private foreign investors to the country, which made it possible for GR to lease Kachung from the state, while many people still lived and farmed the area.

To prepare the land for the GR plantations, the government forcibly evicted many of these people, which led to widespread local resentment against the new plantations. During the evictions, houses, crops, shops and burial sites were destroyed. Local people were vilified as 'encroachers' and 'trespassers' by both government and GR staff (Lyons and Westoby 2014). The combination of these various adverse local livelihood impacts led Lyons and Westoby (2014) to label the GR plantations in Uganda 'carbon colonialism'.

Due to the widespread criticism from researchers and journalists, the SEA decided to pull out of the project in 2015, while Norfund and Finnfund moved in as investors in Green Resources.

In Mozambique, there has recently been a decline in plantation forestry including the GR plantations (Figure 5.1). Among the causes of this decline identified by Mbanze et al. (2022) were increased forest fires set by local people who had provided land for the plantations, but were unhappy with the lack of social benefits. In addition, lack of government commitment and political instability have contributed to the decline in afforestation plantations. These are likely to be common challenges for many afforestation projects in Africa.

THE GREAT GREEN WALL OF AFRICA

The Great Green Wall (GGW) is one of the biggest afforestation projects that are currently being planned in Africa. The project website, a promotional film, and other promotional material, present this project as a 15km wide wall of trees over a stretch of 8000km from Senegal to Djibouti. The project was

Source: Author's own.

Figure 5.1   *Monoculture of pines in a Green Resources plantation in Mozambique*

conceived in 2005 by a group of African Heads of State including the former president of Senegal, Abdoulaye Wade, the former president of Nigeria, Olesegun Obasanjo, and the former leader of Libya, Muammar Gadhafi. It was approved by the African Union as a Pan-African project in 2007 and has later received financial support from the Food and Agriculture Organization of the United Nations (FAO), the World Bank, the Global Environment Facility (GEF), the European Union, and the International Union for Conservation of Nature (IUCN) (UNCCD 2020). At the climate summit in Paris in December 2015, donors pledged a total of 4 billion USD to the project, but by 2020, the project had received merely 870 million USD of this promised funding (UNCCD 2020).

The aims of the GGW are to by 2030 restore 100 million hectares of degraded land, create 10 million jobs, and sequester 250 million tons of carbon. It is believed that these results will furthermore bring down recruitment to jihadist insurgency and reduce migration from the Sahel to Europe.

Due to the project being championed by former president Wade, Senegal is without doubt the leading country in implementing the GGW where most of the land restoration has taken place (leading some to name it 'the great Senegalese wall'). So far, there has been much less project activity in the other Sahelian countries (Magrin and Mugelé 2020). An evaluation initiated by the United Nations Convention to Combat Desertification (UNCCD) concluded that by 2020 only 4 per cent of the planned area had been afforested. This meagre result is due to a general lack of enthusiasm among both Sahelian governments occupied with more pressing issues, as well as among donors who see the project as too risky amidst the security situation in the region (Mugelé 2018, Magrin and Mugelé 2020). The fact that the Sahel has been greening following increased rainfall in the last few decades (see chapter 2) may have led to additional hesitation among donors. Moreover, survival rates of tree seedlings in the Sahel are low unless they are actively watered by hand. According to Yeo (2018), the survival rate in the GGW plantations in Senegal has been 45 per cent following intensive watering and protection of the seedlings.

The documented greening is also said to have recently changed the approach of the project from tree planting 'to become a mosaic of resilient land use systems with the capacity to adapt to uncertainty and climatic extremes' (UNCCD 2020: 29). However, much of the promotional material on various websites among the sponsors of the project still focusses on a wall of trees to stop the advancement of the Sahara Desert, and, through this afforestation, to sequestrate a large amount of carbon.

Despite the dangerous security situation in the Sahel and the scientific critique of the basic assumptions behind such an afforestation project, the World Bank in particular is still pushing for implementation of the GGW in countries like Mali (World Bank 2023), Burkina Faso and Niger (Turner et al. 2021).

The project activities have so far been dominated by a technical and top-down approach focused on tree planting. The project also suffers from lacking local involvement and participation (Mugelé 2018, Magrin and Mugelé 2020, Turner et al. 2021, Turner et al. 2023). Since the project in Senegal is focused on the zone between 100 mm and 400 mm of annual rainfall, which is marginal for dryland farming, the local population consists primarily of pastoralists.

Pastoralism has, however, not been taken into account in the design, planning and implementation of the GGW (Mugelé 2018, Turner et al. 2021, Turner et al. 2023). In the Ferlo region in Senegal, where a large part of project activities has taken place, and where there is a centuries-long history of pastoral use, the pastoral dependence on access to land is not only neglected by the project, but pastoralism is also seen as an obstacle to the afforestation that project success is measured by. Through totally ignoring the needs of pastoralists and their livestock, the GGW exacerbates existing misrecognition of pastoralists in the form of lack of both formal and discursive recognition. Even a publication largely written to promote the great green wall (Boëtsch et al. 2019) admits that the afforestation encroaches on pastoral land and is in conflict with pastoralism in the Ferlo region of Senegal.

First, this misrecognition manifests in the blocking of pastoral mobility through enclosures of afforested areas. Second, it leads to loss of grazing areas that the pastoral system depends on, and third, the afforested areas and vegetable production within these areas compete with livestock for water. Ironically, this means that the Great Green Wall results in natural resource scarcity for the local population in the Ferlo, the Fulani pastoralists (Mugelé 2018).

The top-down approach of the project is also reinforced by the fact that it is implemented by the Water and Forest Department (le Service des Eaux et Forêts), which, as discussed in chapter 2, is an old colonial institution with a para-military tradition of being an armed forest police. In the central project area in the Ferlo region, Mugelé (2018) found that there were eight state forest agents – none of whom is from the region and none who are Fulani. These foresters are given the task of producing success measured by the number of trees planted and the area afforested (Mugelé 2018).

Traditional leaders and the elected leaders of local communes are also largely neglected by the project. They are not consulted and are merely told to comply with the decisions made by the project foresters pertaining, for instance to the location of areas to be afforested. In addition, the foresters give local people fines for illegal use of tree products and when livestock manage to enter afforestation areas (Mugelé 2018). This is possible because the state formally owns all rural land in the Sahel (except individually titled land which is only a small percentage), and pastoral custom is generally not recognised by state legislation or policies (see more above in this chapter). Forest legislation is strict with few local rights of access to forest products and with fines for infraction of rules. This has given state foresters considerable power compared to local populations.

This top-down and technical approach to forest management has a history dating back to the 1930s in the Sahel, as pointed out in chapter 2, and rather than reducing conflicts, it risks further increasing tensions between pastoralists and state institutions. This tension is behind the recruitment of pastoralists to armed groups labelled 'jihadist' (see chapters 3 and 4) and may therefore, in the long run, contribute to the opposite result of what the project intends.

The misrecognition of Sahelian pastoralists following this project, as well as generally, is both formal in terms of lack of rights as well as discursive. The GGW is, to a large extent, conceived as a climate mitigation project, which led to massive donor interest at COP21 in Paris in 2015 as well as in later COPs (Glasgow and Dubai in particular). The resulting misframing has made the victims of the initiative invisible, for instance on the website of the GGW and in other project presentations.

Since 2015, France, under the leadership of President Macron, has continued to invest funds and prestige in this project. At COP 26 in Glasgow in 2021, Macron hosted a re-launch of the Great Green Wall of Africa together with (then) Prince Charles. Included in the panel of speakers was also Amazon owner Jeff Bezos, who promised one Billion USD to afforestation and forest conservation, including funds to the GGW. Supposedly, all three actors hope to use this project to build or boost a green image without any risks of becoming unpopular among voters, citizens or consumers, none of whom are found among affected populations.

When there is such straightforward misrecognition of the local population, in this case pastoralists in the Sahel, there is a clear risk that climate mitigation through afforestation may lead not only to a failed climate project but also to adverse results such as increased local natural resource scarcity and increased resistance to the state, which might ultimately exacerbate conflict levels (Cavanagh and Benjaminsen 2022).

THE SPECTACULAR FAILURE AND FETISHISATION OF AFFORESTATION IN AFRICA

Afforestation projects are typically suffering from problems of 'leakage' and 'permanence'. Leakage refers to the risk that deforestation is displaced outside the project area, while permanence refers to the risk of stored carbon being released through processes such as fire, disease, pests, or human encroachment.

These risks make carbon an 'uncooperative commodity', especially in forest-related climate mitigation projects (Bumpus 2011). Moreover, carbon's lack of cooperation is associated with the complicated and contested processes of creating baselines and justifying additionality. This complexity leads to 'a constant tension between the international carbon market and local socionatural relations' (Bumpus 2011: 624).

In the case of the Green Resources plantations, there were no reliable measurements of a baseline, making the additionality of these plantations questionable. For instance, vegetation has, as a general practice, been cleared before the seedlings of pine and eucalyptus have been planted. In addition, parts of the plantations in Tanzania have also been set on fire by local people several times as a clear expression of resistance against the project.

This illustrates the vulnerability of afforestation as a climate change mitigation instrument and its problem of permanence. This also implies a problem of carbon accounting, since trees that have been destroyed in a fire may already have been sold on a carbon market and accounted for in a carbon budget.

In addition, monocultures of pine and eucalyptus clearly also have negative effects on biological diversity as well as numerous adverse livelihood implications, as shown in the case of the Green Resources plantations.

These problems are, however, generally hidden to Northern consumers or governments who buy carbon credits from afforestation projects – unless critical journalists or researchers expose these problems publicly. Lyons and Westoby (2014) label this information gap between production of forest carbon and consumption as 'carbon fetishism'.

Marx used the term 'commodity fetishism' to highlight that there is often an ideological veil obscuring our understanding of capitalistic production – meaning that commodities tend to appear as objects detached from the social relations they are produced under. This gap widens the more global the commodity becomes. The consumer therefore knows little about the social and environmental conditions of production of particular commodities (Benjaminsen and Svarstad 2021).

When carbon fetishism is exposed, however, a 'spectacular failure' may follow, unraveling mediatised depictions of triple-win scenarios often presented in websites, project documents, and social media. This reveals the spectacular gap between 'representation' and 'execution' in project activities, and the ways in which this gap entails damaging consequences for local communities and ecosystems (Cavanagh and Benjaminsen 2014).

This gap between how a project is sold in websites, project documents and social media and what happens on the ground becomes bigger the more global the commodity is and the longer the distance is between production and consumption. In this sense, carbon is an ideal commodity to fetishise where the consequences at the production site are hidden to consumers.

INTRODUCING CONTEXT

Carbon fetishism also implies that one should take quantifications of stored carbon through afforestation and forest conservation with a pinch of salt. Again, this is an example where context may take 'revenge' (Olivier de Sardan 2021a, see more below).

Quantitative methods and statistics have a long history of domination in the field of practical politics and policy formulation in all sectors of society, also related to environment, development and climate issues. While these approaches have value and contribute to knowledge production in various ways depending on the topic of investigation and the quality of the quantitative data, which, by the way is often questionable, there is a blatant blind spot in quantitative studies related to local, or even regional, socionatures. Or, put differently, there is generally a blind spot in these studies of historical, political and environmental context.

Governments and politicians are generally more convinced by quantitative data than what is often seen as 'anecdotal' evidence that cannot be generalised beyond the single case. These claims about qualitative and case-based studies being merely anecdotal are quite common among large-N researchers as well as politicians and some large Non-Governmental Organisations (NGOs). This is despite the fact that case studies are well established as an approach that does not focus on statistical testing of facts, but rather aims to generate ideas for theory building and analytical generalisation (as opposed to statistical generalisation) (Ragin and Becker 1992, Lund 2014).

Quantification is also seen as more convenient by states as it may be more immediately applicable without disturbing contextual questions of history and power being raised. Contextual analyses, on the other hand, are usually seen by states to complicate state governance as it tends to depend on 'simplification' and 'legibility' in order to rule efficiently (Scott 1998).

Quantitative analyses often play a key role in these processes of simplification and making legible. For instance, simplification and standardisation of pastoral landscapes and practices through measures such as carrying capacity, carcass weight and boundary making form part of the state's attempts at making society 'legible' (Johnsen et al. 2015). Furthermore, Li (2007) argues that in the art of governing, the state needs to establish a serious problem that its policy will solve. In the case of pastoralism, this will often take the form of 'overgrazing', economic 'inefficiency', or increased land-use conflicts. The state may then claim that this problem can only be solved through quantifiable scientific and technical means. Li (2007) calls these two steps 'problematization' and 'rendering technical', and quantitative science plays a key role in both steps (Johnsen et al. 2015).

Moreover, Berkes (2008) argues that the positivist-reductionist approach that has dominated Western science has shaped contemporary environmental governance into depending on 'value-free' generalisations independent of context, space and time. The implication is that problems that are rendered technical by government policy and practice are simultaneously rendered non-political.

Controlled experiments and 'randomised control trials' (Duflo et al. 2007) represent recent additions to quantitative approaches to development studies that have become influential in the last 10–15 years among international aid organisations such as the World Bank and Norad. In 2019, three economists at the Massachusetts Institute of Technology (MIT) received the Nobel Prize in economics for their work with randomised control trials related to aid and development. Their approach is inspired by medical research in laboratories where results from one sample that has received a certain treatment are compared to a sample that has not received this treatment.

Norad has, for instance, engaged this research group at MIT to conduct randomized control trials to investigate how and why measures to arrest deforestation in the Democratic Republic of Congo work. It is, however, easy to imagine how aspects that are not readily quantifiable are neglected or put aside and thereby not included in the analysis. A common critique against forest conservation projects is that there is a tendency to reproduce injustice or even increase inequality and injustice. These methods will most likely exclude questions of injustice or simply be unable to include such questions due to the mere nature of the methods themselves.

Political ecology and critical agrarian studies stand in contrast to these context-less methodologies, being focused on the political, economic, historical and environmental context of socioenvironmental processes. It is, for instance hard to see how large-N studies, experiments or randomized trials can help understand the complex background to the jihadist rebellion in the Sahel.

In his book 'La revanche des contextes: Des mésaventures de l'ingénierie sociale, en Afrique et au-delà' French anthropologist Jean-Pierre Olivier de Sardan asks why there is often a gap between the aims of development projects on the one hand and their actual outcomes on the other (Olivier de Sardan 2021a). He answers this question by pointing at the lack of knowledge about context that 'takes revenge' on projects. What is often presented as unexpected consequences could have been predicted if context were considered.

Olivier de Sardan is also critical of randomised control trials:

- First, while this method is marketed as the only sound scientific approach to impact studies within development aid projects, it remains an ideal tool for context-less standardisation.

- Second, (ironically) the method implies generalisation to a whole population from only one experiment.
- Third, this is an expensive method but it often leads to banal self-evident results.

In addition, we may add that this is not a method that helps to question power imbalances or issues of injustice or inequality. There is a limited set of questions that can be addressed through this method. These are often more practical questions such as if school fees are introduced – what will the impact of school attendance be? (It will go down.)

The science behind quantitative 'value-free' generalisations presents itself as objective, neutral and apolitical, while in practice it tends to be based on specific norms and values with a considerable impact on the research conclusions. Hence, such apolitical ecologies are inherently political, and political ecology is not more 'political' than these approaches. It is just more explicit about its normative goals (Robbins 2012).

While quantitative value- and context-free science continues to dominate in terms of its influence on policy-making, there is also an opposite process developing in parallel. Globally, there is an increasing recognition of Indigenous Knowledge and Local Knowledge (IK & LK) reflected among others, in the recent reports by the Intergovernmental Panel on Climate Change (IPCC) and the Intergovernmental Science-Policy Platform on Biodiversity and Ecosystem Services (IPBES) (Ara Begum et al. 2022, IPBES 2019). For instance, IPBES (2019: XXXVII) mentions that:

> At least one quarter of the global land area is traditionally managed, owned, used or occupied by indigenous peoples. These areas include approximately 35 per cent of the area that is formally protected, and approximately 35 per cent of all remaining terrestrial areas with very low human intervention ... Community-based conservation institutions and local governance regimes have often been effective, at times even more effective than formally established protected areas, in preventing habitat loss ... Several studies have highlighted contributions by indigenous peoples and local communities in limiting deforestation, as well as initiatives showing synergies between these different mechanisms.

Likewise, the IPCC states:

> Indigenous knowledge refers to the understandings, skills and philosophies developed by societies with long histories of interaction with their natural surroundings. Local knowledge is defined as the understandings and skills developed by individuals and populations, specific to the places where they live. These definitions relate to the debates on the world's cultural diversity, which are increasingly connected to climate change debates. However, there is agreement that, in the same way that there is not a unique definition of Indigenous Peoples because it depends

on self-determination (see below), there is not a single definition of neither IK and LK. Therefore, contextualisation is greatly needed. IK and LK will shape perceptions which are vital to managing climate risk in day-to-day activities and longer-term actions. … Such experience-based and practical knowledge is obtained over generations through observing and working directly within various environments. Knowledge may be place based and rooted in local cultures, especially when it reflects the beliefs of long-settled communities who have strong ties to their natural environments (Ara Begum et al. 2022: 148).

This growing international recognition of IK & LK comes together with an increased focus on climate justice reflected in the Sixth Assessment Report of the IPCC (Ara Begum et al. 2022) as well as in a rising global climate justice movement (Newell et al. 2021, see also chapter 1). These discursive trends also relate to a changing global view on pastoralism. In a report for UNEP and IUCN, McGahey et al. (2014) argue, for instance that pastoralism is one of the planet's most sustainable food production systems that may go well together with a transformation to a 'green economy'.

Yet, 'recognition' is not a magic wand that will necessarily change access to land and resources in practice. As German (2022) demonstrates, there is often a gap between formal recognition and gaining access to land in reality or that the knowledge imbued in pastoral practices is taken seriously by state agencies. Through an example from Mozambique, German illustrates how formal state recognition may not substantially enhance the security of customary rights holders with collective tenures. For instance, as shown by Benjaminsen et al. (2009) formalisation of land rights may lead to exclusion of some historical land users, which also brings us back to Honneth's distinction between legal and intersubjective recognition (chapter 1).

In addition to legal state recognition not being sufficient to ensure the realisation of access and benefits, a conception of recognition that depends on the state or other entities from 'above' has also been criticised from a decolonial perspective (Pulido and de Lara (2018). Hence, just as 'participation' may be seen as a top-down concept (Cooke and Kothari 2001), the same critique may be valid for a state-centric view on 'recognition'. While communities depend on this form of recognition to increase their status and power, it may also be a double-edged sword with risks of cooptation, simplification and misrecognition in practice.

Forms of recognition also depend on presentation and representation. Decolonising recognition would depend on listening to the 'senses of justice' among subaltern groups (Svarstad and Benjaminsen 2020). This is, however, not straightforward. As an outsider, one cannot distribute questionnaires or carry out interviews and expect to be able to capture genuine local voices.

James Scott (1985, 1990) famously argues in his work on 'everyday resistance' that marginalised social groups perform differently 'offstage' and 'onstage'. Onstage or 'public transcripts' consist of conversations and statements that the actors play out in what they perceive to be official contexts when they are not sure of anonymity or of how the information will be used. In such cases, it makes sense to play safe and not say anything that may be perceived as controversial. Offstage presentations, on the other hand, or 'hidden transcripts', are accounts that subordinate actors communicate in the absence of the powerful, and they reflect conversations among these groups and typically include the subjects' critique of power and its practices.

This is why distributing questionnaires or carrying out interviews without proper introduction or spending time in communities will not be able to capture the hidden transcripts. Understanding such contextual information takes time and depends on the building of trust.

Political ecology has often critiqued policy-making on environmental issues, but without any significant political impact. This sort of critique focused on context and power relations does generally not fit the agenda of the bureaucrats or politicians in charge. As Walker put it some time ago, there seems to be a mutual antipathy between political ecologists and policy-makers (Walker 2006). Contextual analyses that raise fundamental questions about the premises behind policies and practices are often considered irrelevant and thus excluded from having any political impact. Decision-makers instead demand instrumental knowledge that they can immediately use, and that does not threaten their own position.

There may therefore be limited opportunities for contextual analyses to be directly useful. In a longer time perspective, however, one may hope that such scholarly contributions may contribute to influencing the general public and environmental and climate activists and thereby contribute to changing perspectives and ultimately political agendas.

NOTES

1. Afforestation implies the conversion to forest of land that has historically not contained forest.
2. This presentation of degrowth relies partly on Hickel's arguments in a recent debate with Stéphane Hallegatte, lead economist in the World Bank (Hickel and Hallegatte, 2021) and partly on his contribution to another debate – in the journal *Political Geography* (Hickel 2021).
3. Jevons Paradox says that with technological improvements and higher efficiency in resource use costs go down, which may again increase demand leading to rising resource use. This principle was first described by the English economist William Stanley Jevons in 1865.
4. BECCS is still a speculative methodology and 'requires the production of bioenergy feedstocks (generally fast-growing species like miscanthus or switchgrass) which must be transported to where they will be converted to steam/

heat, liquid fuels, or charcoals. These fuel products are then used for energy generation and the emitted CO_2 is captured either pre- or post-combustion and stored' (McElwee 2023: 192).

5. A/R = Afforestation and reforestation.
6. According to FAO (2004) a forest plantation has 'few species, even spacing and/or even-aged stands'.
7. GR had first an intentional agreement with the Norwegian Ministry of Finance that would buy the carbon credits from the plantations in Tanzania. The Ministry, however, withdrew when GR did not succeed in obtaining CDM certification for these plantations.

6.   Climate security and climate justice: taking context and science seriously

This book has focused on issues of climate security and climate justice in the Sahel, which are two traditions of social science-based environmental and climate research that are rarely brought into dialogue. In particular, there is a gap in the literature on the interface (both overlaps and tensions) between climate security and climate justice that this book tries to fill. Both these approaches are confronted with context in this book.

To further develop the idea of 'context', the approach takes inspiration from political philosophers, in particular Nancy Fraser and Axel Honneth, who have written extensively on 'recognition' as a key aspect of justice. In addition, the book draws on neo-Marxist approaches within peasant and critical agrarian studies – taking particular inspiration from seminal research in the 1960s and 70s that attempted to explain the emergence of peasant rebellions. For instance, E.P. Thompson saw peasant resistance in England in the 19th century as a response to a perceived threat to a 'moral economy' and a 'subsistence ethic' by capitalist development leading to land dispossession, or Primitive Accumulation as Marx called it (which David Harvey has later re-named Accumulation by Dispossession). James Scott further developed this idea in his book on 'Moral Economy of the Peasant' (1976) focussing on peasant rebellions in South-East Asia, while Eric Wolf in his 'Peasant Wars in the Twentieth Century' (1969) compared the emergence of eight peasant rebellions in the Global South through a similar Marxian lens with Primitive Accumulation being a common denominator in these rebellions.

While this approach to understanding the context of peasant rebellion has been widely applied to analyse uprisings in Latin America and parts of Asia, it is largely absent from studies of grassroots uprisings in Africa, and in particular in attempts to understand the background to the emergence of so-called 'jihadist' armed groups in the Sahel.

This book combines such a historical materialist approach within critical agrarian studies and political ecology with environmental discourse analysis. As shown in several chapters, discursive presentations through rhetorical devices such as narratives and images of Sahelian societies and landscapes since early in the colonial period until the present have worked to dispossess people from their land and to put restrictions on their use of natural resources. In other words, discourses matter and have material consequences for people's lives.

We see this in the enduring discourse on desertification resulting in attempts to arrest this alleged process through large-scale afforestation or through restrictions on local use. We also see it in dominant media and policy ideas about climate-driven conflicts in the Sahel that tend to a-politicise conflicts and to divert attention from historical and political causes as well as bringing back old tropes about climate determinism in Africa. And we see the effect of discourses on people's lives through widespread views on pastoralism as an outdated form of production that is said to be economically inefficient, ecologically destructive and geographically expansive leading to land-use conflicts. This lack of recognition resulting in the continued loss of pastures and livestock corridors over several decades has contributed to the current largely pastoral uprising in the Sahel.

All this adds up to a number of external historical constructions about the Sahel and its people that have had direct implications for the current crisis in the region. Even 'the Sahel' itself seems to be a construction without roots in local geographical conceptions.

On the other hand, there are also internal tropes in the Sahel about, for instance, Fulani as equivalent to jihadists in addition to the development of a new geopolitical anti-French, anti-Western and pro-Russian discourse where Putin is celebrated as an anti-colonialist. Curiously, both the current military regimes in the three Sahelian countries Mali, Burkina Faso and Niger, as well as the jihadist groups they fight against, have their own anti-colonial agendas.

Jihadist groups on the one hand aim to replace schools where French is the language of instruction with Quranic schools where the curriculum consists of one book only, which is in Arabic. Books are seen to be 'haram' because they are said to promote Western decadence, undermine Islam and traditional values and prop up a national corrupt elite.

After France was thrown out of the Sahel, leading to the subsequent reduction of Western development aid and diplomacy, military regimes and jihadist groups fight for control over land and resources in Mali, Burkina Faso and Niger.

In the army, and among a majority of the urban population, anti-French and anti-Western attitudes are common, while democracy is associated with elite rule and corruption, and former elected presidents are regarded as marionettes who were controlled by France.

In addition, a mixture of nationalism and Pan-Africanism has become dominant, but a Pan-Africanism that only includes solidarity among the three mentioned countries (Olivier de Sardan 2024), due primarily to economic sanctions from Ecowas.

While the military regimes enjoy support among the urban populations, various jihadist groups continue to control large rural areas of the three countries. These groups also reject Western democracy, that they associate with a national corrupt elite influenced by Western interests. Leaders of jihadist groups such as Iyad ag Ghaly and Hamadoun Kouffa insist on the introduction of Sharia as a condition for peace, although few fighters seem to have joined these groups for religious reasons. Many pastoralists (Fulani and Tuareg in particular) have joined the rebellion because of widespread grievances related to a corrupt state bureaucracy and what they see as an anti-pastoral policy that has led to continuous loss of dry season pastures and access to livestock corridors over several decades.

This means that decolonisation in practice has definitely arrived in the Sahel. But this is a decolonisation far from any academic debate. It is not epistemology, which is the concern of people in Bamako, Ouagadougou or Niamey who demonstrate in favour of military regimes, shout 'down with France' and celebrate Putin as an anti-colonial hero. Olivier de Sardan (2024) has labelled this process 'military decolonization'. In addition, however, there is also the competing 'jihadi decolonization' project that should not be forgotten.

While neither the military regimes, nor people in general in the Sahel, nor jihadists highlight climate change as an important factor behind this overwhelming security crisis, large parts of the international media and the international aid system as well as some researchers continue to focus on climate change as a key factor contributing to this security crisis.

CLIMATE SECURITY

Climate change became an international security issue from the early 2000s following the debate on 'environmental security' that had emerged in the 1990s. This securitisation of climate change first took place within international politics and policy-formulation. In contrast to other aspects of international debates on climate change, which have been driven by science, climate security has first and foremost been a politically driven agenda. The political attention has furthermore led to a global media attention to the issue as well as making financial resources available for climate action and research. This increased attention, including the availability of funding, has led NGOs, development agencies, recipient governments and researchers to explore this issue from various angles.

Some researchers have aimed to develop a climate security approach, often without much empirical substance, while quantitative researchers have pointed at the lack of connection between climate change variables and the emergence of conflicts, and qualitative researchers have highlighted the lack of historical and political contextual analyses in climate security contributions and that in order to understand conflict and crisis one needs to study such contexts.

This question about the role of climate change in creating conflicts seems, however, to be never-ending and non-resolvable as demonstrating lack of historical connection or correlation does not impress those who are convinced about such a connection and who instead tend to use the future as empirical evidence – 'if climate wars haven't arrived as yet, they certainly will in the future!'

The problem with this perspective is that it depoliticises and dehistoricises conflicts and reproduces external and rather colonial views of the Sahel without recognising Sahelian historical and political contexts. This may be by convenience, as understanding history and politics and taking local actors seriously demands some effort. At the same time, a simplistic climate security narrative fits the agenda of a number of actors from international institutions, Northern governments (who see an opportunity of greenwashing their policies), NGOs and climate activists as well as Sahelian governments.

As discussed previously (chapters 1, 3, 4 and 5), emphasising the role of climate change without recognising context easily leads to reproduction of Eurocentrism, environmental determinism, and green neo-colonialism. Despite good intentions, climate security risks reflecting these effects unless there is more explicit engagement with context, empirical knowledge and the politics of climate justice. Hence, there is a need to decolonise climate security in the Sahel through the involvement of Sahelian history, politics and agency in the analysis.

We have seen this persistent coloniality in, for instance, the negligence of context in climate security analyses as well as in afforestation projects such as the Great Green Wall.

Climate security research has, however, recently moved into the issue of climate risks more broadly, presumably as a response to the above-mentioned impasse in climate security research relating climate variables directly or through threat multipliers to the emergence of conflicts. The focus is currently rather on how climate change-related risks may occur via possible indirect pathways as, for instance through reduced food security or the implementation of adaptation or mitigation measures. Some of this research highlights state capacity and the level of socio-economic development as generally more important factors than climate change in explaining conflict, while also stressing that climate change-related risks may potentially impact on conflicts. In addition, conflict risks associated with climate mitigation initiatives such as afforestation, windmills, and biofuels are also increasingly discussed in this recent literature on climate change-related risks.

Some climate security analysts have also commendably begun to criticise the climate security literature for not paying enough attention to causal mechanisms (Busby 2022). Exploring possible causal mechanisms, Busby did a comparative case-study analysis of seven country cases in Africa and Asia. The conclusion was that there are three factors in particular that tend to lead climate change to trigger conflict and humanitarian emergencies: 1) Weak state capacity; 2) exclusive political institutions; and 3) foreign assistance that is blocked or delivered unevenly.

These three points may, however, be debated.

First, it is often said that African, and in particular Sahelian, states are weak and fragile (Craven-Matthews and Englebert 2018, Bøås and Strazzari 2020, Osland and Erstad 2020). The policy response to this widely shared analysis has been development programmes aimed at capacity-building to strengthen state governance.

However, people on the ground that have had encounters with the Sahelian state may have experienced this state as anything but weak. When armed for-esters in green uniforms with off-road motorbikes give women fines for col-lecting dry dead wood, the experience of these women is that this is a state with power and the potential to use force through violence. This force has been backed up by a Forest Service that has received wide international financial and moral support to stop desertification.

The same goes for pastoralists who have lost access to key resources or spaces because of government decisions. They may not have experienced the state as weak. In other words, people who have been at the sharp end of bureaucratic power (e.g. women who collect wood, pastoralists or small-scale farmers who have lost access to land through elite capture by state officials, or shop owners who get visits from customs officials collecting bribes) do not necessarily experience the Sahelian state as weak. The rebellion can rather be seen as a reaction against the abuse of state power.

Viewing Sahelian states as weak or fragile implies assessing these states within a Westphalian framework where the Western state is the model they are compared to. In contrast to this view, Olivier de Sardan (2021b) argues that the Sahelian states were brutally created basically from nothing by colonial pow-ers. These new states became 'despotic in nature' and far from the model of the French state, and they were continued after independence developing their own bureaucratic culture and practice of elite capture. The Sahelian state is not more traditional than European states, argues Olivier de Sardan. It is just mod-ern in a different way. The combination of top-down governance, bureaucratic corruption and lack of provision of services has over time led to a lack of trust in the state among its citizens.

This means that the Sahelian state has followed its own trajectory. It is therefore not a 'failed' Western-style state, but another type of state – a rentier state. This means that while some traditional state functions are weak in the

Sahel (health services, education), other sectors are strongly governed (land, natural resources).

The second causal mechanism according to Busby, exclusive political institutions, is more convincing. He defines political inclusion as 'incorporation of all politically and militarily relevant subgroups in government decision-making and "fair" apportionment of resources and programs" (Busby 2022: 46). Political exclusion is exactly what pastoralists in particular, but also small-scale farmers in the Sahel have been complaining about for decades and which the jihadists have acted on. Much of this book is devoted to this theme.

The third point, the role of foreign assistance, is more complex. Busby's overall assumption here, while recognising that aid may play several roles, 'is that international assistance can compensate for state weakness in the lead up to, and in the wake of, hazard exposure' (Busby 2022: 55).

If the analysis is that conflicts and rebellion are generally caused by poverty and joblessness, this may be a fair assumption, since foreign assistance presumably provides resources and jobs to people. But such an assumption and analysis, which stand in contrast to the argument in this book, also underestimates the political reasons for people joining armed groups (see chapter 3 and 4). This is, however, a common and often repeated claim in international debates – that people join jihadist groups because of poverty, and the money they are paid when joining, and that aid and the creation of jobs represent a main solution to this problem. But most pastoralists who join armed groups do not look for employed work, they join because of development practices, from governments and international donors, that have drained pastoralism as a production system of its resources. If livestock corridors are not blocked and if dryland pastures are not converted to rice fields, most pastoralists will make much more money from livestock keeping than from any manual job they would be able to get.

In the Sahel, much development aid has contributed to the rentier state rather than being its correction, although many smaller NGOs have provided essential humanitarian aid as well as capacity building for its staff. In Mali, international development assistance has contributed to the rentier state through, for instance large-scale investments such as Office Riz Mopti (chapter 3) and the Office du Niger (chapter 5) as well as how a militarised Forest Service was propped up by international support to stop desertification in the 1980s and 90s (chapter 2). The subsequent abuse of force and corruption led to widespread grievances among rural people that jihadist leaders can still draw on as arguments in their speeches highlighting corruption among state elites.

In other words, it is not difficult to find examples of foreign assistance in the Sahel that has contributed to poor state governance, elite capture, more conflicts and to the overall current crisis. More of this type of context-less assistance ignores how potential adverse effects will lead to more, not less, climate insecurity.

CLIMATE JUSTICE

Among climate activists, academics and policy-makers there are different interpretations of what climate justice implies. All actors, however, share a concern for fairness, equity, and justice in the context of climate change.

The term itself has become well-established in international climate negotiations (Conference of the Parties – COP) as well as in the recent reports by the IPCC. Distributive justice combined with the historical responsibilities of the Global North, which is recognised through the principle 'common but differentiated responsibilities' remains the main focus of most southern governments in addition to many climate activists.

The urgency of the climate crisis also requires immediate global action to bring about social transformations towards low-carbon societies, which again implies deep structural societal change with widespread implications for social justice.

Ecomodernism has been the overwhelming approach followed by rich countries to aim for such a transformation. This approach aims to decouple economies from environmental impacts through the reliance on markets and on technological development. Ecomodernism also has a focus on developing intensified large-scale high-tech agriculture combined with urbanisation and the conservation of 'wilderness'. The implication is that the Global South should urbanise and replace small-scale farming and pastoralism with large technology-intensive farms that can feed urban populations.

This also means that most people should live in cities, while rural areas are used for industrial farming and the conservation of 'unused' areas. In other words, ecomodernists do not seem to be overly concerned about the practical implications for climate justice of these proposals as they will necessarily have enormous consequences for rural peoples' livelihoods following widespread land dispossession. Also, UN agencies including the IPCC are dominated by ecomodernist thinking as exemplified by climate scenarios that are based on the use of BECCS and afforestation on areas corresponding to 2–3 times the Indian land mass (see chapter 5).

Degrowth presents itself in international climate and environmental debates as the main contrasting alternative to ecomodernism. It implies a reduction of resource and energy use in high-income countries to stay within planetary boundaries. Degrowth advocates point out that the current overshoot of planetary boundaries is primarily driven by rich countries and that the flip side of a continued ecomodernist approach in these countries based on the transition to renewable energy such as solar and wind power necessarily will be increased extractivism in the Global South, where most of the materials for these technologies come from.

The most sustainable – and painful – solution to the challenge of how to transform to a low-carbon society therefore inevitably involves radical reductions in levels of production and consumption in rich countries. Thus far, this remains, however, a debate among activists and academics and has had meagre implications on political debates within nation states or international organisations.

Many climate activists, NGOs, and climate scientists also tend to neglect the implications for African land-use and land dispossession of a continued eco-modernist approach in the Global North. It is, on the contrary, rather common to frame climate justice as an international obligation to invest in climate mitigation measures in Africa such as the Great Green Wall or other afforestation projects. According to a number of NGOs and climate activists – the Sahara Desert is advancing towards the south, and this is an issue of climate justice.

This thinking, however, leads to neo-colonial climate solutions in the form of afforestation, often on pastoral land, including pastoral dispossession and potentially increased conflict levels. The implication is that many organisations and individuals who think they work for climate justice actually do the opposite due to their lack of knowledge of context – and the result of their proposals may therefore be more climate injustice.

Such apolitical proposals tend to neglect the insight that vulnerability to climate change has political and historical causes linked to exploitation and marginalisation. In other words, vulnerability does not fall from the sky (Ribot 2010). It should be studied within a wider political ecology framework.

It should be clear from the previous chapters in this book that the coloniality of knowledge production in the Sahel remains. Wilkens and Datchoua-Tirvauday (2022) argue that decolonising climate justice demands questioning the distribution of resources and burden-sharing, participation in the creation of rules in climate governance, and questioning whose knowledge, concepts and ways of knowing are recognised.

> … inherited colonial mindsets and (neo)colonial practices still prevail in the academic interdisciplinary field of climate research, and in global climate governance design and decision-making. This leads, on the one hand, to the replication of coloniality of research practices in academia and in the ways of disseminating research to policy-makers, states and non-state actors; and on the other hand, to knowledge hierarchies shaping the consultation processes in global climate governance, the discourses on climate change and the choice of climate policies, often neglecting the diverse ways of knowing of those most affected by climate change (Wilkens and Datchoua-Tirvauday 2022: 127).

'Agrarian climate justice' has emerged from critical agrarian studies as an alternative view on climate justice in the rural South that takes issues of power relations, access to land, coloniality and recognition of local histories and practices seriously.

Recognition is a key aspect of justice, and therefore also of climate justice, which is often neglected in mainstream approaches. This book has aimed to demonstrate how recognition placed in a historical, political and environmental context may facilitate a deeper understanding of the social processes behind rising insecurity and social-environmental processes more broadly.

CONTEXT, POLITICS AND SCIENCE

The book has also argued that a good comprehension of context related to climate security and climate justice depends on qualitative research to understand social dynamics, power relations and historical trajectories. But it also depends on sound natural science to assess landscape and ecological change over time as well as past and future climate change.

While quantitative methods dominate practical politics and policy-formulation, these methods tend to neglect historical, political and environmental context. Governments and politicians are, however, more convinced by quantitative data than case studies and qualitative methods that are often framed as 'anecdotal' evidence that cannot be generalised beyond the single case. Such views are common among large-N researchers as well as 'practitioners' (government and NGO people), despite the fact that case studies do not focus on the statistical testing of facts but rather aim to generate ideas for theory-building and analytical generalisation (as opposed to statistical generalisation).

Quantification is also seen as apolitical and innocent and therefore more useful by governments as it may be more immediately applicable without raising inconvenient questions about power relations. Contextual analyses, on the other hand, are seen to complicate state governance.

Quantitative analyses also often play a key role in processes of simplification and standardization of, for instance, pastoral landscapes and practices through measures such as carrying capacity, carcass weight and boundary making. In the art of governing, the state also needs to establish a serious problem that its policy will solve. In the case of pastoralism, this will often take the form of 'overgrazing', economic 'inefficiency', or increased land-use conflicts. The state may then claim that this problem can only be solved through quantifiable scientific and technical means.

In the field of climate politics in the Sahel, many of these policy-led processes with their paraphernalia of World Bank and NGO projects, consultancies, and applied research that are often focused on the primacy of quantifiable data tend, ironically, to also ignore the research frontier in ecology and climate science.

A recent example is the World Bank project 'Restoration of degraded land in Mali', which surprisingly started in 2023 in the midst of a national and regional security crisis. The budget of this project of 150 million USD is a loan, which is even more surprising – why should the government of Mali borrow that amount of money to plant trees to fight desertification when public services in the country are largely closed down due to the security situation? And why should the World Bank be willing to lend this type of money under this situation of widespread violence and political uncertainty? Who are responsible for such a careless and extravagant waste of funds?

But leaving these questions aside, the project is said to be the World Bank's contribution to the implementation of the Great Green Wall and it aims to contribute to climate change mitigation through tree planting on savanna land and thereby to stop desertification, which is said to be caused by overgrazing, overcutting of wood for fuel and other local human activities in addition to climate change.

Moreover, the project evaluation document (World Bank 2023) also states that the project is based on the understanding that there is a close link between degradation of natural resources, climate change and conflicts:

> For example, it has been shown that in the Sahelian context, with a change by 1 per cent in temperature and rainfall, there is a corresponding increase in the frequency of trans-frontier violence of 4 per cent, and simultaneously, inter-communal violence may increase up to 14 per cent (World Bank 2023: 13) (translated from French by the author).

This document is indeed reminiscent of the anti-desertification euphoria of the 1980s and 90s, only this time under even more puzzling conditions due to the national security crisis in Mali, the political instability in the country and the availability of a large amount of research, both from the social and natural sciences, going directly against the ideas that the project is based on.

Recent studies in ecology demonstrate, for instance, that grasslands and savannas have a high capacity to store carbon below ground and that planting trees in these places may jeopardise that capacity. In a similar vein, dryland ecosystems can often be described as non-equilibrial systems where setting a carrying capacity does not make sense due to the stochasticity of the system, meaning that overgrazing also becomes a problematic term. In addition, recent climate models do not predict less total rainfall and desertification in the Sahel, but rather higher quantities of rain, but also probably more variability. This means that the challenge for people in the Sahel is not drier conditions and desertification, but how to adapt to increasing variability.

For pastoralists who have centuries of experience in coping with climatic variability and in tracking the resulting fluctuations in the availability of pastures, politics and not climate change represent the main obstacle to their way of life and the sustainability of their production systems.

Dryland farmers, on the other hand, will experience increased rainfall variability as increased uncertainty, potentially leading to declining agricultural production and food security. There are, however, other market and state-related dependencies that make these farmers poor and vulnerable in the first place. Secure land tenure and less extraction from rent-seeking elites are examples of factors that would help decrease vulnerability. Government and aid policies that actually support small-scale farmers and pastoralists on their terms, instead of being based on Eurocentric ideas of desertification and local overuse of natural resources, would also be helpful. Instead of being based on a poor contextual understanding and neglecting scientific facts, these policies should recognise the logic and knowledge inherent in local production systems based on a long history of working the land and its resources.

The recent World Bank project in Mali, as part of the Great Green Wall in Africa, demonstrates, however, the robustness and resilience of Eurocentric and neo-colonial narratives about the Sahel. This pessimistic insight is corroborated by the fact that scientific facts and knowledge of context tend to lose faced with the power of politics and money that are largely driven by political interests in sustaining ecomodernist approaches to climate and environmental governance in the rich countries of the North.

Bibliography

Abrahams, D. and E. Carr. 2017. Understanding the Connections Between Climate Change and Conflict: Contributions from Geography and Political Ecology. *Current Climate Change Reports* 3/4.

Adger, W.N., T.A. Benjaminsen, K. Brown and H. Svarstad. 2001. Advancing a political ecology of global environmental discourses. *Development and Change* 32 (4): 681–715.

Ag Baye, C. 1993. The process of a peace agreement. Between the movements and the United Fronts of Azawad and the government of Mali. In Veber, H., J. Dahl, F. Wilson and E. Wæhle (eds): *...Never drink from the same cup*. Proceedings of the conference on indigenous peoples in Africa. Tune, Denmark, 1993. IWGIA Document no. 74: pp. 247–56.

Al Jazeera. 2019. Best way to fight climate change? Plant a trillion trees. 4th July. https://www.aljazeera.com/news/2019/7/4/best-way-to-fight-climate-change-plant-a-trillion-trees#:~:text=Study%20says%20over%20decades%2C%20new,of%20heat%2Dtrapping%20carbon%20dioxide.&text=The%20most%20effective%20way%20to,enough%20room%2C%20Swiss%20scientists%20say

Amselle, J.L. 2022. *L'invention Du Sahel*. Paris: Éditions du Croquant.

Anglo-French Commission. 1973. Mission forestière Anglo-Francaise Nigeria-Niger (Décembre 1936-Févier 1937). *Bois et Forêts des Tropiques* 148: 3–26.

Angus, I. 2023. *The war against the commons. Dispossession and resistance in the making of capitalism*. New York: Monthly Review Press.

Anyamba, A. and C.J. Tucker. 2005. Analysis of Sahelian vegetation dynamics using NOAA-AVHRR NDVI data from 1981–2003. *Journal of Arid Environment* 63 (3): 596–614.

Ara Begum, R., R. Lempert, E. Ali, T.A. Benjaminsen, T. Bernauer, W. Cramer, X. Cui, K. Mach, G. Nagy, N.C. Stenseth, R. Sukumar and P. Wester. 2022. Point of Departure and Key Concepts. In: Climate Change 2022: Impacts, Adaptation and Vulnerability. Contribution of Working Group II to the Sixth Assessment Report of the Intergovernmental Panel on Climate Change [H.-O. Pörtner, D.C. Roberts, M. Tignor, E.S. Poloczanska, K. Mintenbeck, A. Alegría, M. Craig, S. Langsdorf, S. Löschke, V. Möller, A. Okem, B. Rama (eds)]. Cambridge University Press, Cambridge, UK and New York, NY, USA: pp. 121–96.

Asafu-Adjaye, J., L. Blomquist, S. Brand, B. Brook, R. Defries, E. Ellis, C. Foreman, et al. 2015. *An ecomodernist manifesto*. Oakland: The Breakthrough Institute.

Asiyanbi, A. and J.F. Lund. 2020. Policy Persistence: REDD+ Between Stabilization and Contestation. *Journal of Political Ecology* 27(1): 378–400.

Assouma, M.H., et al. 2019. Contrasted seasonal balances in a Sahelian pastoral ecosystem result in a neutral annual carbon balance. *Journal of Arid Environments* 162: 62–73.

Aubréville, A. 1938. Assèchement et déforestation de l'A.O.F. In *La forêt coloniale.* Soc. d'Ed. Géogr. Maritimes et Coloniales, Paris: 25–42.

Aubréville, A. 1949. *Climats, forêts et désertification de l'Afrique tropicale.* Paris: Société d'Editions Géographiques, Maritimes et Coloniales.

Ba, B. 2008. *Pouvoirs, ressources et développement dans le delta central du Niger.* Paris: L'Harmattan.

Ba, B. 2016. *Crises de gouvernance: Justice transitionnelle et paix durable au Mali.* Bamako: La Sahélienne.

Ba, B. and S. Cold-Ravnkilde. 2021. When Jihadists Broker Peace: Natural Resource Conflicts as Weapons of War in Mali's Protracted Crisis. DIIS Policy Brief January 2021. Danish Institute for International Studies (DIIS).

Bächler, G. 1998. *Violence through environmental discrimination: Causes, Rwanda arena, and conflict Model.* Dordrecht: Kluwer Academic.

Barrière, O. and C. Barrière. 2002. *Un droit à inventer. Foncier et environnement dans le delta intérieur du Niger. Editions IRD, collection: A travers champs.* Paris: Institut de Recherche pour le Développement (IRD).

Barthes, R. 1961 (1977). *Image, music, text.* New York: Hill & Wang.

Barthes, R. 1972 (2013). *Mythologies.* New York: Hill & Wang.

Bassett, T.J. 1988. The political ecology of peasant-herder conflicts in the northern Ivory Coast. *Annals of the Association of American Geographers* 78 (3): 453–72.

Bassett, T.J. 1995. The uncaptured corvée: Cotton in Côte d'Ivoire, 1912–1946. In: Isaacman, A., Roberts, R. (eds), *Cotton, Colonialism and Social History in Sub-Saharan Africa.* Portsmouth: Heinemann/James Currey.

Bassett, T.J. 2001. *The Peasant Cotton Revolution in West Africa. Côte d'Ivoire, 1880–1995.* Cambridge University Press, Cambridge.

Bassett, T.J. and K. Bi Zuéli. 2000. Environmental discourses and the Ivorian savanna. *Annals of the Association of American Geographers* 90 (1): 67–95.

Bastin, J.F., Y. Finegold, C. Garcia, D. Mollicone, M. Rezende, D. Routh, C.M. Zohner and T.W. Crowther. 2019. The global tree restoration potential. *Science* 365: 76–9.

Baudais, V. 2015. *Les trajectoires de l'état au Mali.* Paris: L'Harmattan.

Bayart, J.F. 1993. *The state in Africa: politics of the belly.* Harlow: Longman.

Bayart, J.F., S. Ellis and B. Hibou. 1999. *The criminalization of the state in Africa.* Oxford: James Currey.

Becker, L.C. 2001. Seeing green in Mali's woods: colonial legacy, forest use, and local control. *Annals of the Association of American Geographers* 91 (3): 504–26.

Beek, van, W. 2005. The Dogon heartland: Rural transformations on the Bandiagara escarpment. In: De Bruijn, M., H. Van Dijk, M. Kaag and K. Van Til (eds) *Sahelian Pathways: Climate and Society in Central and South Mali.* Report 78/2005: pp. 40–70. Leiden: African Studies Centre.

Behnke, R.H. 2000. Equilibrium and non-equilibrium models of livestock population dynamics in pastoral Africa: Their relevance to Arctic grazing systems. *Rangifer* 20 (2–3): 141–52.

Behnke R.H., I. Scoones and C. Kerven. 1993. Range ecology at disequilibrium: new models of natural variability and pastoral adaptation in African savannas. Overseas Development Institute (ODI), London, UK.

Benjaminsen, T.A. 1993. Fuelwood and desertification: Sahel orthodoxies discussed on the basis of field data from the Gourma region in Mali. *Geoforum* 24 (4): 397–409.

Benjaminsen, T.A. 1996. Bois-énergie, déboisement et sécheresse au Sahel. Le cas du Gourma malien. *Sécheresse* 7 (3): 179–85.

Benjaminsen, T.A. 1997. Natural resource management, paradigm shifts, and the decentralization reform in Mali. *Human Ecology* 25 (1): 121–43.

Benjaminsen, T.A. 2000. Conservation in the Sahel: Policies and people in Mali (1900-1998). In: Broch-Due, V. and R. Schroeder (eds). *Producing nature and poverty in Africa*. Uppsala: Nordic Africa Institute: pp. 94–108.

Benjaminsen, T.A. 2001. The population-agriculture-environment nexus in the Malian cotton zone. *Global Environmental Change* 11 (4): 27–39.

Benjaminsen, T.A. 2008. Does Supply-Induced Scarcity Drive Violent Conflicts in the African Sahel? The Case of the Tuareg Rebellion in Northern Mali. *Journal of Peace Research* 45 (6): 819–36.

Benjaminsen, T.A. 2015. Political Ecologies of Environmental Degradation and Marginalization. In *The Routledge Handbook of Political Ecology*, edited by T. Perreault, G. Bridge, and J. McCarthy: pp. 354–65. London: Routledge.

Benjaminsen, T.A. 2023. The risks of ecological security. *New Perspectives*, 31 (1).

Benjaminsen, T.A. and B. Ba. 2009. Farmer-herder conflicts, pastoral marginalisation and corruption: a case study from the inland Niger delta of Mali. *The Geographical Journal* 174 (1): 71–81.

Benjaminsen, T.A. and B. Ba. 2019. Why do pastoralists in Mali join jihadist groups? A political-ecological explanation. *Journal of Peasant Studies* 46 (1): 1–20.

Benjaminsen, T.A. and B. Ba. 2021. Fulani-Dogon Killings in Mali: Farmer-Herder Conflicts as Insurgency and Counterinsurgency. *African Security* 14 (1): 4–26.

Benjaminsen, T.A. and G. Berge. 2004a. Myths of Timbuktu – From African El Dorado to Desertification. *International Journal of Political Economy* 34 (1): 31–59.

Benjaminsen, T.A. and G. Berge. 2004b. Histoires de Tombouctou. Arles. Actes Sud.

Benjaminsen, T.A. and I. Bryceson. 2012. Conservation, green/blue grabbing and accumulation by dispossession in Tanzania. *Journal of Peasant Studies* 39 (2): 335–55.

Benjaminsen, T.A. and P. Hiernaux. 2019. From Desiccation to Global Climate Change: A History of the Desertification Narrative in the West African Sahel, 1900–2018. *Global Environment* 12 (1): 206–36.

Benjaminsen, T.A. and C. Lund. 2002. Formalisation and informalisation of land and water rights in Africa: An introduction. *European Journal of Development Research* 14 (2): 1–10.

Benjaminsen T.A. and H. Svarstad 2021. *Political Ecology: A Critical Engagement with Global Environmental Issues*. London: Palgrave-Macmillan.

Benjaminsen, T.A., J. Aune and D. Sidibé. 2010. A critical political ecology of cotton and soil fertility in Mali. *Geoforum* 41 (4): 647–56.

Benjaminsen, T.A, H. Svarstad and I. Shaw of Tordarroch. 2022. Recognizing recognition in climate justice. *IDS Bulletin* 53 (4).

Benjaminsen, T.A., K. Alinon, H. Buhaug and J. T. Buseth. 2012. Does climate change drive land-use conflicts in the Sahel? *Journal of Peace Research* 49 (1): 97–111.

Benjaminsen, T.A., S. Holden, C. Lund and E. Sjaastad. 2009. Formalisation of land rights: Some empirical evidence from Mali, Niger and South Africa. *Land Use Policy* 26: 28–35.

Benjaminsen, T.A., H. Reinert, E. Sjaastad and M.N. Sara. 2015. Misreading the Arctic landscape: A political ecology of reindeer, carrying capacities and 'overstocking' in Finnmark, Norway. *Norwegian Journal of Geography* 69 (4): 219–29.

Benjaminsen, T.A., Rohde, R.F., Sjaastad, E., Wisborg, P. and Lebert, T. 2006. Land reform, range ecology, and carrying capacities In Namaqualand, South Africa. *Annals of the Association of American Geographers* 96 (3): 524–40.

Berge, G. 1999. In Defence of Pastoralism: Form and Flux Among Tuaregs in Northern Mali. Thesis Report Series no. 1 2000. PhD dissertation. SUM-University of Oslo.

Berge, G. 2001. Tuareg notions of space and place in Northern Mali. In Benjaminsen, T.A. and Lund, C. (eds) *Politics, Property and Production: Understanding Natural Resources Management in the West African Sahel*. Uppsala: Nordic Africa Institute.

Bergius, M. and J.T. Buseth. 2019. Towards a green modernization development discourse: the new green revolution in Africa. *Journal of Political Ecology* 26 (1).

Bergius, M., T.A. Benjaminsen and M. Widgren. 2018. Green Economy, Scandinavian Investments and Agricultural Modernization in Tanzania. *Journal of Peasant Studies* 45 (4): 825–52.

Bergius, M., T.A. Benjaminsen, F. Maganga and H. Buhaug. 2020. Green Economy, Degradation Narratives and Land-Use Conflicts in Tanzania. *World Development* 104850.

Berkes, F. 2008. *Sacred Ecology*. New York: Routledge.

Bertrand, A., C. Cossalter and D. Laurent. 1984. Planification de l'énergie. Etude du secteur des combustibles forestiers au Mali (Bois de feu et charbon de bois). Paris: CTFT/TransEnerg.

Beusekom, van, M.M. 1999. From underpopulation to overpopulation: French perceptions of population, environment, and agricultural development in French Soudan (Mali), 1900–1960. *Environmental History* 4: 198–219.

Biasutti, M. 2013. Forced Sahel rainfall trends in the CMIP5 archive. *Journal of Geophysical Research: Atmospheres* 118: 1613–1623.

Blaikie, P. and H. Brookfield (eds). 1987. *Land degradation and society*. London: Methuen.

Bluwstein, J. and C. Cavanagh. 2023. Rescaling the land rush? Global political ecologies of land use and cover change in key scenario archetypes for achieving the 1.5°C Paris agreement target. *Journal of Peasant Studies* 50 (1), 262–94.

Bøås, M. 2015. Crime, coping, and resistance in the Mali-Sahel periphery. *African Security* 8: 299–319.

Bøås, M. and F. Strazzari. 2020. Governance, fragility and insurgency in the Sahel: A hybrid political order in the making. *The International Spectator*, 55 (4), 1–17.

Bøås, M. and L.E. Torheim. 2013. The Trouble in Mali—Corruption, Collusion, Resistance. *Third World Quarterly* 34 (7): 1279–92.

Boeke, S. 2016. Al Qaeda in the Islamic Maghreb: Terrorism, Insurgency, or Organized Crime? *Small Wars & Insurgencies* 27 (5): 914–36.

Boeke, S. and B. Schuurman. 2015. Operation 'Serval': A Strategic Analysis of the French Intervention in Mali, 2013–2014. *The Journal of Strategic Studies* 38 (6): 801–25.

Boeke, S. and A. Tisseron. 2014. Mali's Long Road Ahead. *The RUSI Journal* 159 (5): 32–40.

Boëtsch, G., P. Duboz, A. Guisse and P. Sarr. 2019. *La grande muraille verte. Une réponse africaine au changement climatique*. Paris: CNRS Editions.

Boiley, P. 1999. *Les Touaregs Kel Adagh. Dépendances et révoltes: du Soudan français au Mali contemporain*. Paris: Karthala.

Boisvert, M.A. 2015. Failing at violence: The longer-lasting impact of pro-government militias in northern Mali since 2012. *African Security* 8: 272–98.

Bond, W.J., N. Stevens, G.F. Midgley and C.E.R. Lehmann. 2019. The Trouble with Trees: Afforestation Plans for Africa. *Trends in Ecology and Evolution* 34 (11): 963–65.

Borras, S.M. and J.C. Franco. 2018. The challenge of locating land-based climate change mitigation and adaptation politics within a social justice perspective: towards an idea of agrarian climate justice. *Third World Quarterly* 39 (7): 1308–25.

Borrel, T., A. Boukari-Yabara, B. Collombat and T. Deltombe (eds). 2021. *L'Empire qui ne veut pas mourir: une histoire de la Françafrique*. Paris: Seuil.

Boudet G. 1972. Désertification de l'Afrique tropicale sèche. *Adansonia*, sér. 2, 12 (4): 505–24.

Bourgeot, A. 2019. Dans le marigot malien. Malinet. Accessed [16.4.2024]. https://www.maliweb.net/contributions/andre-bourgeot-dans-le-marigot-malien-2831665.html

Brandt, M., P. Hiernaux, K. Rasmussen, C. Mbow, L. Kergoat, T. Tagesson, Y.Z. Ibrahim, A. Wélé, J.C. Tucker, R. Fensholt. 2016. Assessing woody vegetation trends in Sahelian drylands using MODIS based seasonal metrics. *Remote Sensing of Environment* 183: 215–25.

Brandt, M., A. Verger, A.A. Diouf, F. Baret and C. Samimi. 2014. Local vegetation trends in the Sahel of Mali and Senegal using long term time series FAPAR satellite products and field measurement (1982–2010). *Remote Sensing* 6: 2408–34.

Brandts, A. 2005. Coping strategies of Dogon cultivators of the northern escarpment. In: De Bruijn, M., H. Van Dijk, M. Kaag and K. Van Til (eds) *Sahelian Pathways: Climate and Society in Central and South Mali*. Report 78/2005: pp. 71–94. Leiden: African Studies Centre.

Brinkerhoff, D.W., and J.D. Gage. 1993. *Forestry policy reform in Mali: An analysis of implementation issues*. USAID, Washington.

Briske, D.D., S. Vetter, C. Coetsee, and M.D. Turner. 2024. Rangeland Afforestation Is Not a Natural Climate Solution. *Frontiers in Ecology and the Environment*, March.

Brockington, D. 2002. *Fortress conservation: The preservation of the Mkomazi Game Reserve, Tanzania*. Oxford: James Currey.

Brown, O., A. Hammil and R. McLeman. 2007. Climate change as the 'new' security threat: implications for Africa. *International Affairs* 83 (6): 1141–54.

Buhaug, H. 2010. Warming not to blame for African civil wars. *PNAS* 107 (38): 16477–82.

Buhaug, H., T.A. Benjaminsen, E.A. Gilmore and C. Hendrix. 2023. Climate-driven risks to peace over the 21st century. *Climate Risk Management* 39: 100471.

Bumpus, A.G., 2011. The matter of carbon: understanding the materiality of tCO2e in carbon offsets. *Antipode* 43 (3): 612–38.

Busby, Joshua. 2022. *States and Nature: The Effects of Climate Change on Security*. Cambridge: Cambridge University Press.

Büscher, B. 2009. Letters of gold: enabling primitive accumulation through neoliberal conservation. *Human Geography* 2 (3): 91–4.

Büscher, B. 2016. Nature 2.0: Exploring and theorizing the links between new media and nature conservation. *New Media & Society* 18 (5): 726–743.

Buxton, N. and B. Hayes. 2015. *The Secure and the Dispossessed: How the Military and Corporations are Shaping a Climate-Changed World*. London: Pluto Press.

Calmon, D., Jacovetti, C. and Koné, M. 2021. Agrarian climate justice as a progressive alternative to climate security: Mali at the intersection of natural resource conflicts. *Third World Quarterly* 42 (12): 2785–803.

Castelli, L. de. 2014. Mali: From Sanctuary to Islamic State. *The RUSI Journal* 159 (3): 62–8.

Cavanagh, C. 2018. Enclosure, Dispossession, and the Green Economy: New Contours of Internal Displacement in Liberia and Sierra Leone? *African Geographical Review* 37 (2): 120–33.

Cavanagh, C. 2021. Limits to (de)Growth: Theorizing 'the Dialectics of Hatchet and Seed' in Emergent Socio-Ecological Transformations. *Political Geography* 90: 102479.

Cavanagh, C. and T.A. Benjaminsen. 2014. Virtual nature, violent accumulation: The 'spectacular failure' of carbon offsetting at a Ugandan National Park. *Geoforum* 56: 55–65.

Cavanagh, C. and T.A. Benjaminsen 2017. Political ecology, variegated green economies, and the foreclosure of alternative sustainabilities. *Journal of Political Ecology* 24: 200–16.

Cavanagh, C. and T.A. Benjaminsen. 2022. Conservation, Land Dispossession and Resistance in Africa. In *Oxford Handbook of Land Politics*.

Chandler, D. 2017. Semiotics for beginners. http://www.visual-memory.co.uk/daniel/ Documents/S4B/

Charbonneau, B. 2019. Intervention as counter-insurgency politics. *Conflict, Security & Development* 19 (3): 309–14.

Charbonneau, B. 2020. Sahel: la gouvernance contre-insurrectionnelle. *Bulletin Francopaix* 5 (1).

Chayanov, A. 1925 (1966). *The theory of peasant economy*. Manchester: Manchester University Press.

Chudeau, R. 1921. Le problème du desséchement en Afrique Occidentale. *Bulletin du Comité d'Etudes Historiques et Scientifiques de L'AOF*: 353–69.

Cissé, M.G. 2018. Hamadoun Koufa, fer de lance du radicalism dans le Mali central. In: De Bruijn, M. (ed) *Biographies de la Radicalisation: Des messages cachés du changement social*. Bamenda, Cameroon: pp. 181–202, Langaa RPCIG.

Clements F.E., 1916. *Plant succession: an analysis of the development of vegetation*. Carnegie Inst. Wash. Publ. 242: 1–512.

Cline, L.E. 2013. Nomads, islamists, and soldiers: The struggles for Northern Mali. *Studies in Conflict & Terrorism* 36 (8): 617–34.

Cline-Cole, R.A., J.A. Falola, H.A.C. Main, M.J. Mortimore, J.E. Nichol, and E.D. O'Reilly. 1990. *Woodfuel in Kano*. Tokyo: The United Nations University Press.

Comolli, V. 2015. *Boko Haram: Nigeria's islamist insurgency*. Oxford University Press.

Cooke, B. and U. Kothari (eds). 2001. *Participation: The New Tyranny?* London: Zed Books.

Corson, C. 2011. Territorialization, enclosure and neoliberalism: non-state influence in struggles over Madagascar's forests. *Journal of Peasant Studies* 38 (4): 703–26.

Cotula, L. and S. Cissé. 2006. Changes in 'customary' resource tenure systems in the inner Niger delta, Mali. *Journal of Legal Pluralism and Unofficial Law* 52: 1–29.

Craven-Matthews, C. and P. Englebert. 2018. A Potemkin state in the Sahel? The empirical and the fictional in Malian state reconstruction. *African Security* 11 (1): 1–31.

Cristiani, D. and R. Fabiani. 2013. The Malian Crisis and its Actors. *The International Spectator* 48 (3): 78–97.

D'Alisa, G., F. Demaria and G. Kallis (eds) 2014. *Degrowth: a vocabulary for a new era*. London: Routledge.

Dalby, Simon. 2022. *Rethinking Environmental Security*. Edward Elgar Publishing.

Daniel, S. 2012. *AQMI. L'industrie de l'enlèvement*. Mesnil-sur-l'Estrée: Fayard.

Davies, J. and M. Nori. 2008. Managing and mitigating climate change through pastoralism. *Policy Matters*, 16, October 2008.

Davis, D. 2016a. *The Arid Lands: History, Power, Knowledge*. Cambridge, Mass.: MIT Press.

Davis, D. 2016b. Deserts and drylands before the age of desertification, in Behnke, R.H. and Mortimore M. (eds). *The end of desertification? Disputing environmental change in the drylands*. Heidelberg: Springer Earth System Sciences.

Day, A. and F. Krampe. 2023. Beyond the UN Security Council: Can the UN General Assembly tackle the climate–security challenge? Stockholm International Peace Research Institute (SIPRI), commentary https://www.sipri.org/commentary/essay/2023/beyond-un-security-council-can-un-general-assembly-tackle-climate-security-challenge

De Bruijn, M. and J. Both. 2017. Youth between state and rebel (dis)orders: Contesting legitimacy from below in Sub-Saharan Africa. *Small Wars & Insurgencies* 28 (4–5): 779–98.

De Bruijn, M. and H. van Dijk. 1995. *Arid ways. Cultural understanding of insecurity in Fulbe society, central Mali*. Amsterdam: Thela Publishers.

De Bruijn, M. and H. Van Dijk. 2005. Moving people: Pathways of Fulbe pastoralists in the Hayre-Seeno area, Central Mali. In: De Bruijn, M., H. Van Dijk, M. Kaag and K. Van Til (eds) *Sahelian Pathways: Climate and Society in Central and South Mali*. Report 78/2005: pp. 247–73. Leiden: African Studies Centre.

De Gironcourt, G. 1910. Sur la productivité des pays Soudanais du Moyen-Niger. *Bulletin De La Société De Géographie Commerciale De Paris* 32: 251–362.

De Gironcourt, G. 1912. Le sommet de la boucle de Niger: Géographie physique et botanique. *Bulletin De La Société De Géographie* 25 (3): 153–71.

Deprez, A. et al. 2024. Sustainability limits needed for CO_2 removal. *Science* 383: 484–6.

Desgrais, N., Y. Guichaoua and A. Lebovich. 2018. Unity is the exception. Alliance formation and de-formation among armed actors in Mali. *Small Wars & Insurgencies* 29 (4): 654–79.

Detzi, D. and S. Winkleman. 2016. Hitting them where it hurts: A joint interagency.

Diallo, O.A. 2017. Ethnic clashes, jihad, and insecurity in central Mali. *Peace Review: A Journal of Social Justice* 29: 299–306.

Dijk, van, H. and M. de Bruijn. 1995. Pastoralists, chiefs and bureaucrats: A grazing scheme in dryland central Mali. In: van den Breemer, J.P.M., C.A. Drijver and L.B. Venema (eds). *Local resource management in Africa*: pp. 77–95. Chichester: John Wiley & Sons.

Direction nationale des eaux et forêts (DNEF), République du Mali (1985). Plan national de lutte contre la désertification et l'avancée du désert, 1985–2000. Bamako.

Dong, S. 2016: *Overview: Pastoralism in the World*, 1–37. ISBN 978-3319307305.

Dowd, C. 2015. Grievances, governance and Islamist violence in sub–Saharan Africa. *Journal of Modern African Studies* 53 (4): 505–31.

Dowd, C. and C. Raleigh. 2013. The myth of global Islamic terrorism and local conflict in Mali and the Sahel. *African Affairs* 112 (448): 498–509.

Dryzek, J.S. 1997. *The politics of the earth. Environmental discourses*. Oxford: Oxford University Press.

Duflo, E., R. Glennerster and M. Kremer. 2007. Using Randomization in Development Economics Research: A Toolkit. *Handbook of Development Economics* Volume 4: 3895–962.

Dyer, G. 2010. *Climate wars. The fight for survival as the world overheats.* Oxford: One world.

Eckholm, E.R., G. Foley, G. Barnard and L. Timberlake. 1984. *Firewood: The Energy Crisis that Won't Go Away.* London: Earthscan.

Edelman, M. and W. Wolford. 2017. Introduction: Critical agrarian studies in theory and practice. *Antipode* 49 (4): 959–76.

Eizenga, D. and W. Williams. 2020. Le puzzle formé par le JNIM et les groupes islamistes militants au Sahel. *Bulletin de la Sécurité Africaine* No. 38.

Eklundh, L., and L. Olsson. 2003. Vegetation index trends for the African Sahel 1982– 1999. *Geophysical Research Letters* 30: 1430–3.

Fairhead, J. 2001. International dimensions of conflict over natural and environmental resources. In Peluso, N.L. and M. Watts (eds.) *Violent environments*: pp. 213–36. Ithaca: Cornell University Press.

Fairhead, J. and M. Leach. 2000. Desiccation and domination: science and struggles over environment and development in colonial Guinea. *Journal of African History* 41: 35–54.

Fairhead, J., M. Leach and I. Scoones. 2012. Green grabbing: a new appropriation of nature? *Journal of Peasant Studies* 39 (2): 237–61.

FAO. 2004. Global forest resources assessment update 2005. Terms and definitions. Forest Resources Assessment WP 83.

Federici, S. 2004. *Caliban and the witch. Women, the body and primitive accumulation.* Milton Keynes: Penguin.

Fensholt, R., I. Sandholt, S. Stisen and C. Tucker. 2006. Analysing NDVI for the African continent using the geostationary meteosat second generation SEVIRI sensor. *Remote Sensing of Environment* 101 (2): 212–29.

Filipovich, J. 2001. Destined to Fail: Forced Settlement at the Office du Niger, 1926– 45. *The Journal of African History* 42 (2): 239–60.

Fisher, S. 2015. The emerging geographies of climate justice. *Geographical Journal* 181 (1): 73–82.

Fok, M., 1994. Le développement du coton au Mali par analyse des contradictions. Les acteurs et les crises de 1895 à 1993. Unité de Recherche, *Economie des Filières,* no 8. Montpellier: CIRAD.

Forsyth, T. 2003. *Critical political ecology. The politics of environmental science.* London: Routledge.

Fraser, J. 2018. Amazonian struggles for recognition. *Transactions of the Institute of British Geography* 43: 718–32.

Fraser, N. 1998. Social justice in the age of identity politics: Redistribution, recognition and participation. In Ray, L. and A. Sayer (eds). *Culture and economy after the cultural turn.* London: Sage.

Fraser, N. 2000. Rethinking Recognition. *New Left Review* 3.3: 107–18.

Fraser, N. 2009. *Scales of Justice: Reimagining Political Space in a Globalizing World.* New York NY: Columbia University Press.

Fraser, N. and Honneth, A. 2003. *Redistribution or Recognition? A Political-Philosophical Exchange.* London: Verso.

Gallais, J. 1975. *Pasteurs et Paysans du Gourma.* Paris: CNRS.

Galy, M. (ed.). 2013. *La guerre au Mali: Comprendre la crise au Sahel et au Sahara. Enjeux et zones d'ombre.* Paris: La Découverte.

Gautier, D., T.A. Benjaminsen, L. Gazull and M. Antona. 2013. Neoliberal forest reform in Mali: Adverse effects of a World Bank 'success'. *Society and Natural Resources* 26: 702–16.

German, L. 2022. *Power, knowledge, land: Contested ontologies of land and its governance in Africa.* Ann Arbor: University of Michigan Press.

Giannini, A. 2016. 40 years of climate modelling: The causes of late-20th century drought in the Sahel. In: R. Behnke and M. Mortimore (eds). *The end of desertification? Disputing environmental change in the drylands.* Heidelberg: Springer.

Giraud, G. 2013. Cinquante ans de tensions dans la sahélo-saharienne. In Galy, M. (ed) *La guerre au Mali. Comprendre la crise au Sahel et au Sahara, Enjeux et zones d'ombre.* Paris: La Découverte.

Gleditsch, N.P. 1998. Armed conflict and the environment: a critique of the literature. *Journal of Peace Research* 35 (3): 363–80.

Gleditsch, N.P. and R. Nordås. 2014. Conflicting messages? The IPCC on conflict and human security. *Political Geography* 43: 82–90.

Goldman, M.J., P. Nadasdy and M. Turner (eds). 2011. *Knowing nature: conversations at the intersection of political ecology and science studies.* Chicago, IL: University of Chicago Press.

Gonin, P., N. Kotlok and M.-A. Pérouse de Montclos (eds). 2013. *La tragédie malienne.* Paris: Vendémiaire.

Grove, R. 1997. *Ecology, climate and empire: colonialism and global environmental history, 1400–1940.* Cambridge: White Horse Press.

Grubler, A., C. Wilson, N. Bento, B. Boza-Kiss, V. Krey, D. L. McCollum, … J. Cullen. 2018. A low Energy Demand Scenario for Meeting the 1.5°C Target and Sustainable Development Goals Without Negative Emission Technologies. *Nature Energy* 3 (6): 515–27.

Guichard, F., Kergoat, L., Hourdin, F., Léauthaud, C., Barbier, J., Mougin, E. and Diarra, B. 2015: Le réchauffement climatique observé depuis 1950 au Sahel. in *Evolutions récentes et futures du climat en Afrique de l'Ouest: Evidences, incertitudes et perceptions* Lalou R. and Sultan B. (eds), IRD Editions, 1–13.

Gupta, D.K. 2007. The Naxalites and the Maoist Movement in India: Birth, Demise, and Reincarnation. *Democracy and Security* 3 (2): 157–88.

Hajdu, F., O. Penje and K. Fischer. 2016. Questioning the use of 'degradation' in climate mitigation: A case study of a forest carbon CDM project in Uganda. *Land Use Policy* 59: 412–22.

Hajer, M. 1995. *The politics of environmental discourse.* Oxford: Oxford University Press.

Hanne, O. (ed) 2014. *Mali, une paix à gagner. Analyses et témoignages sur l'opération Serval.* Panazol: Lavauzelle.

Hansen, S.J. 2019. *Horn, Sahel and Rift. Fault-lines of the African Jihad.* London: Hurst & Co.

Harmon, S.A. 2014. *Terror and insurgency in the Sahara-Sahel region.* London: Routledge.

Hartmann, B. 2014. Converging on disaster: Climate security and the Malthusian anticipatory regime for Africa. *Geopolitics* 19: 757–83.

Harvey, D. 2003. *The New Imperialism.* Oxford: Oxford University Press.

Hatløy, A. 1999. Methodological Aspects of Assessing Nutrition Security: Examples from Mali. Ph.D. dissertation, Institute for Nutrition Research, University of Oslo.

Herrmann, S.M., A. Anyamba and C.J. Tucker, 2005: Recent trends in vegetation dynamics in the African Sahel and their relationship to climate. *Global Environmental Change* 15 (4): 394–404.

Hickel, J. 2021. The anti-colonial politics of degrowth. *Political Geography* 88 (3): 102404.

Hickel, J. and S. Hallegatte. 2021. Can We Live within Environmental Limits and Still Reduce Poverty? Degrowth or Decoupling? *Development Policy Review* 40 (1).

Hiernaux, P. 1988. Vegetation monitoring by remote sensing: Progress in calibrating a radiometric index and its application in the Gourma, Mali. *ILCA Bulletin* 32: 14–21.

Hiernaux, P., C. Dardel, L. Kergoat and E. Mougin 2016. Desertification, adaptation and resilience in the Sahel: Lessons from long term monitoring of agro-ecosystems. In Behnke, R.H. and M. Mortimore (eds). *The end of desertification? Disputing environmental change in the drylands*. Heidelberg: Springer.

Higazi, A., B. Kendhammer, K. Mohammed, M.A. Pérouse de Montclos, and A. Thurston. 2018. A Response to Jacob Zenn on Boko Haram and al-Qa'ida. *Perspectives on Terrorism* 12 (2): 203–13.

Holder, G. 2023. À propos des Peuls « qui ne sont pas nés » : esclavage, djihad et droits humains au centre du Mali. *Afriquecontemporaine* 2023/2 (276), 221–44.

Homer-Dixon, T. 1994. Environmental scarcities and violent conflict. Evidence from cases. *International security* 19 (1): 5–40.

Homer-Dixon, T. 1999. *Environment, scarcity, and violence*. Princeton: Princeton University Press.

Honneth, A. 1995. *The struggle for recognition: The moral grammar of social conflicts*. Cambridge, MA: MIT Press.

Honneth, A. 2001. Recognition or redistribution? Changing perspectives on the moral order of society. *Theory, Culture, and Society* 18 (2–3): 43–55.

Hubert, H. 1917. Progression du dessèchement dans les régions sénégalaises. *Annales de géographie* 26: 376–85.

Hubert, H. 1920. Le dessèchement progressif en Afrique Occidentale. *Bulletin du comité d'études historiques et scientifiques de l'Afrique Occidentale Francaise*: 401–67.

Huckabey, J.M. 2013. Al Qaeda in Mali: The defection connections. *Orbis* 57 (3): 467–84.

Hulme, M. 2014. Reducing the future to climate: a story of climate determinism and reductionism. *Osiris* 26 (1): 245–66.

Hussein, K., J. Sumberg and D. Seddon. 1999. Increasing Violent Conflict between Herders and Farmers in Africa: Claims and Evidence. *Development Policy Review* 17: 397–418.

Hutchinson, C.F., S. M. Herrmann, T. Maukonen and J. Weber. 2005. Introduction: The 'Greening' of the Sahel. *Journal of Arid Environments* 63: 535–7.

Ibrahim, Y.I. 2017. The wave of jihadist insurgency in West Africa: Global ideology, local context, individual motivations. West African Papers, No. 07, OECD Publishing, Paris.

Ibrahim, Y.I. and M. Zapata. 2018. Regions at risk. Preventing mass atrocities in Mali. United States Holocaust Memorial Museum & Simon Skjodt Center for the Prevention of Genocide. Washington D.C.

Igoe, J. 2017. *The Nature of Spectacle: On Images, Money, and Conserving Capitalism*. Tucson: University of Arizona Press.

International Crisis Group, 2020. Enrayer la communautarisation de la violence au centre du Mali. Report Number 293.

IPBES. 2019. Global assessment report on biodiversity and ecosystem services of the Intergovernmental Science-Policy Platform on Biodiversity and Ecosystem

Services. E.S. Brondizio, J. Settele, S. Díaz and H.T. Ngo (eds). IPBES secretariat, Bonn, Germany. 1148 pages.

Jafry, T., Mikulewicz, M. and Helwig, K. 2018. Introduction: Justice in the Era of Climate Change. In *Routledge Handbook of Climate Justice*. London: Routledge.

Johnsen, K.I., T.A. Benjaminsen and I.M.G Eira. 2015. Seeing like the state or like pastoralists? Conflicting narratives on the governance of Sámi reindeer husbandry in Finnmark, Norway. *Norwegian Journal of Geography* 69 (4): 230–41.

Johnson, M. 1976. The economic foundations of an Islamic theocracy: The case of Masina. *Journal of African History*, 17 (4), 481–95.

Jones, S.M. and D.S. Gutzler, 2016. Spatial and Seasonal Variations in Aridification across Southwest North America. *J. Climate* 29 (12): 4637–49.

Jourde, C., M. Brossier and M.G. Cissé. 2019. Prédation et violence au Mali: élites statuaires peules et logiques de domination dans la région de Mopti. *Canadian Journal of African Studies* 53 (3): 431–45.

Kahl, C.H. 2006. *States, scarcity, and civil strife in the developing world*. Princeton: Princeton University Press.

Kalyvas, S.N. 2003. The ontology of 'political violence': Action and identity in civil wars. *Perspectives on Politics* 1 (3): 475–94.

Karumbidza, B. and W. Menne. 2011. CDM Carbon Sink Tree Plantations: A Case Study in Tanzania. The Timberwatch Coalition.

Kashwan, K. and Ribot, J. 2021. Violent Silence: The Erasure of History and Justice in Global Climate Policy. *Current History* 120 (829): 326–31.

Keenan, J. 2013. *The dying Sahara. US imperialism and terror in Africa*. London: Pluto Press.

Kelly, A.B. 2011. Conservation practice as primitive accumulation. *Journal of Peasant Studies* 38 (4): 683–701.

Kipling, R. 1889. Poem: *The Ballad of East and West*. In Rudyard Kipling's Verse (Definitive ed.). 1940. Garden City, NY: Doubleday. pp. 233–6.

Koubi, V. 2019. Climate change and conflict. *Annual Review of Political Science* 22: 343–60.

Kouyaté, S. 2006. Etude des enjeux nationaux de protection du basin du fleuve Niger. Report to Groupe de Coordination des Zones Arides (GCOZA), Bamako.

Krätli, S. and N. Schareika, 2010. Living Off Uncertainty: The Intelligent Animal Production of Dryland Pastoralists. *European Journal of Development Research* 22 (5): 605–22.

Kreibiehl, S., T. Yong Jung, S. Battiston, P.E. Carvajal, C. Clapp, D. Dasgupta, N. Dube, R. Jachnik, K. Morita, N. Samargandi, M. Williams, 2022: Investment and finance. In IPCC, 2022: Climate Change 2022: Mitigation of Climate Change. Contribution of Working Group III to the Sixth Assessment Report of the Intergovernmental Panel on Climate Change [P.R. Shukla, J. Skea, R. Slade, A. Al Khourdajie, R. van Diemen, D. McCollum, M. Pathak, S. Some, P. Vyas, R. Fradera, M. Belkacemi, A. Hasija, G. Lisboa, S. Luz, J. Malley, (eds.)]. Cambridge University Press, Cambridge, UK and New York, NY, USA.

Lacher, W. 2013. Challenging the myth of the drug-terror nexus in the Sahel. West Africa Commission on Drugs Background Paper No 4. Kofi Annan Foundation.

Lahache, M.J. 1907. Le desséchement de l'Afrique française, est-il démontré? *Bulletin de la société de géographie et d'études coloniales de Marseille* 31: 149–85.

Larder, N. 2015. Space for pluralism? Examining the Malibya land grab. *Journal of Peasant Studies* 42 (3–4): 839–58.

Laris, P. 2002. Burning the Seasonal Mosaic: Preventative Burning Strategies in the Wooded Savanna of Southern Mali. *Human Ecology* 30 (2): 155–86.

Lavauden, L. 1927. *Les forêts du Sahara*. Nancy: Berger-Levrault.

Leach, G. and R. Mearns. 1988. *Beyond the Woodfuel Crisis*. London: Earthscan.

Leach, M. and R. Mearns (eds) 1996. *The lie of the land: challenging received wisdom on the African environment*. Oxford: James Currey.

Lecocq, B. 2010. *Disputed desert. Decolonisation, competing nationalisms and Tuareg rebellions in northern Mali*. Leiden: Brill.

Lecocq, B., G. Mann, B. Whitehouse, D. Badi, L. Pelckmans, N. Belalimat, B. Hall and W. Lacher. 2013. One hippopotamus and eight blind analysts: A multivocal analysis of the 2012 political crisis in the divided Republic of Mali. *Review of African Political Economy* 40 (137): 343–57.

Lefebvre, C. 2021. *Des Pays Au Crépuscule*. Paris: Fayard.

Lentz, C. 2006. Land rights and the politics of belonging in Africa: an introduction in Kuba, R. and Lentz, C. (eds) *Land and the politics of belonging in West Africa*. Leiden: Brill, 1–34.

Levien, M. 2022. Regimes of dispossession. In *The Routledge Handbook of Property, Law and Society*, Graham, N., M. Davies and L. Godden (eds). London: Routledge.

Li, T. M. 2007. *The will to improve*. Durham, N. C.: Duke University Press.

Li, T. M. 2010. To make live or die? Rural dispossession and the protection of surplus populations. *Antipode* 41 (1): 66–93.

Lode, K. 1997. The Peace Process in Mali: Oiling the Works? *Security Dialogue* 28 (4): 409–24.

Lounnas, D. 2014. Confronting Al-Qa'ida in the Islamic Maghrib in the Sahel: Algeria and the Malian crisis. *The Journal of North African Studies* 19 (5): 810–27.

Lund, C. 2014. Of What is This a Case?: Analytical Movements in Qualitative Social Science Research. *Human Organization* 73 (3): 224–31.

Lyons, K. and P. Westoby. 2014. Carbon colonialism and the new land grab: Plantation forestry in Uganda and its livelihood impacts. *Journal of Rural Studies* 36: 13–21.

Mach, K.J., C.M. Kraan, W.N. Adger, H. Buhaug, M. Burke, J.D. Fearon, C.B. Field, C.S. Hendrix, J.-F. Maystadt, J. O'Loughlin, P. Roessler, J. Scheffran, K.A. Schultz and N. Von Uexkull. 2019. Climate as a risk for armed conflict. *Nature* 571: 193–7.

Macia, E., J. Allouche, M. Bassimbé Sagna, A. Diallo, G. Boëtsch, A. Guissé, P. Sarr, J.-D. Cesaro and P. Duboz. 2023. The Great Green Wall in Senegal: Questioning the Idea of Acceleration through the Conflicting Temporalities of Politics and Nature among the Sahelian Populations. *Ecology and Society* 28 (1).

Magrin, G. and R. Mugelé. 2020. La boucle de l'Anthropocène au Sahel: nature et sociétés face aux grands projets environnementaux (Grande Muraille Verte, Sauvegarde du lac Tchad). *Belgeo* 3/20.

Mangin, M. 1924. Une mission forestière en Afrique Occidentale Francaise. *La géographie* 42 (4): 449–83, 629–54.

Marin, A., E. Sjaastad, T.A. Benjaminsen, M.N. Sara and J. Borgenvik. 2020. Productivity beyond density: a critique of management models for reindeer pastoralism in Norway. *Pastoralism* volume 10, Article number: 9 (2020).

Marret, J.L. 2008. Al-Qaeda in Islamic Maghreb: A 'glocal' organization. *Studies in Conflict & Terrorism* 31 (6): 541–52.

Marx, K. 1976. *Capital, Volume 1*. London: Penguin Classics.

Mbanze, A.A., S. Wang, J. Mudekwe, C. Ribas Dias and A. Sitoe. 2022. The Rise and Fall of Plantation Forestry in Northern Mozambique. *Trees, Forests and People* 10: 100343.

McDonald, M. 2021. *Ecological Security: Climate Change and the Construction of Security.* Cambridge University Press.

McElwee, P. 2023. Advocating afforestation, betting on BECCS: land-based negative emissions technologies (NETs) and agrarian livelihoods in the global South. *Journal of Peasant Studies* 50 (1): 185–214.

McGahey, D. Davies, J. Hagelberg, N. and R. Ouedraogo. 2014. *Pastoralism and the Green Economy – a natural nexus?* Nairobi: IUCN and UNEP.

Mehta, L. (ed). 2010. *The Limits to Scarcity: Contesting the Politics of Allocation.* London: Earthscan.

Mingay, G.E. 1997. *Parliamentary Enclosure in England.* London: Longman.

Mirzabaev, A., L.C. Stringer, T.A. Benjaminsen, P. Gonzalez, R. Harris, M. Jafari, N. Stevens, C.M. Tirado and S. Zakieldeen, 2022: Cross-Chapter Paper 3: Deserts, Semiarid Areas and Desertification. In: *Climate Change 2022: Impacts, Adaptation and Vulnerability.* Contribution of Working Group II to the Sixth Assessment Report of the Intergovernmental Panel on Climate Change [H.-O. Pörtner, D.C. Roberts, M. Tignor, E.S. Poloczanska, K. Mintenbeck, A. Alegría, M. Craig, S. Langsdorf, S. Löschke, V. Möller, A. Okem, B. Rama (eds.)]. Cambridge University Press, Cambridge, UK and New York, NY, USA, pp. 2195–231.

Moorhead, R.M. 1991. Structural chaos: community and state management of common property in Mali. PhD dissertation, University of Sussex.

Mugelé, R. 2018. La Grande muraille verte: géographie d'une utopie environnementale au Sahel. PhD dissertation, Université Paris 1 Panthéon-Sorbonne.

National Geographic. 2019. 4 July. https://www.nationalgeographic.com/environment/article/how-to-erase-100-years-carbon-emissions-plant-trees

Neumann, R.P. 1998. *Imposing Wilderness. Struggles over Livelihood and Nature Preservation in Africa.* Berkeley: University of California Press.

Newell, P., S. Srivastava, L.O. Naess, G.A. Torres Contreras and R. Price. 2021. Toward transformative climate justice: An emerging research agenda. *Wiley Interdisciplinary Reviews: Climate Change* 12 (6): e733.

Nijenhuis, K. 2009. Reconsidering conflicts over land in the Sahel as conflicts over power. In: Böcker, A., W. Van Rossum and H. Weyers (eds) *Legal Anthropology from the Low Countries.* Amsterdam: Reed Business. 69–101.

Nordås, R. and N.P. Gleditsch. 2007. Climate change and conflict. *Political Geography* 26: 627–38.

Nwankwo, C.F. 2023. The Moral Economy of the Agatu "Massacre": Reterritorializing Farmer-Herder Relations. *Society* 60: 359–71.

Oakland Institute. 2011. Understanding Land Investment Deals in Africa. Country Report: Mali.

OECD. 2012. Green growth and developing countries: A summary for policymakers. Paris: OECD.

Olivier de Sardan, J.P. 2021a. *La revanche des contextes: Des mésaventures de l'ingénierie sociales en Afrique et au-delà.* Paris: Karthala.

Olivier de Sardan J.P. 2021b. The Construction of States and Societies in the Sahel. *The Oxford Handbook of the African Sahel,* Villalón, L.A. (ed). Oxford University Press.

Olivier de Sardan, J.P. 2024. Au Sahel, un décolonialisme militaire. AOC, 29 février 2024, https://aoc.media/analyse/2024/02/28/au-sahel-un-decolonialisme-militaire/#_ftn11.

Olsson, L., L. Eklundh and J. Ardö. 2005. A recent greening of the Sahel–trends, patterns and potential causes. *Journal of Arid Environments* 63: 556–66.

O'Neill, B., M. van Aalst, Z. Zaiton Ibrahim, L. Berrang Ford, S. Bhadwal, H. Buhaug, D. Diaz, K. Frieler, M. Garschagen, A. Magnan, G. Midgley, A. Mirzabaev, A. Thomas and R. Warren, 2022: Key Risks Across Sectors and Regions. In: *Climate Change 2022: Impacts, Adaptation and Vulnerability*. Contribution of Working Group II to the Sixth Assessment Report of the Intergovernmental Panel on Climate Change [H.-O. Pörtner, D. C. Roberts, M. Tignor, E.S. Poloczanska, K. Mintenbeck, A. Alegría, M. Craig, S. Langsdorf, S. Löschke, V. Möller, A. Okem, B. Rama (eds.)]. Cambridge University Press, Cambridge, UK and New York, NY, USA, pp. 2411–538.

Osland, K.M. and H.U. Erstad. 2020. The Fragility Dilemma and Divergent Security Complexes in the Sahel. *The International Spectator* 55 (4).

Panthou, G., T. Vischel and T. Lebel. 2014. Recent Trends in the Regime of Extreme Rainfall in the Central Sahel. *International Journal of Climatology* 34 (15): 3998–4006.

Parashar, S. 2019. Colonial legacies, armed revolts and state violence: the Maoist movement in India. *Third World Quarterly* 40 (1): 337–54.

Parenti, C. 2011. *Tropic of chaos. Climate change and the new geography of violence*. New York: Nation Books.

Parr, C.L., M. Te Beest and N. Stevens. 2024. Conflation of Reforestation with Restoration Is Widespread. *Science* 383 (6684): 698–701.

Pedersen, J. and T.A. Benjaminsen. 2008. One leg or two? Food security and pastoralism in the northern Sahel. *Human Ecology* 36 (1): 43–57.

Peet, R. and M. Watts. (eds) 1996. *Liberation Ecologies. Environment, Development, Social Movements*. London: Routledge.

Peluso, N.L. and M. Watts (eds). 2001. *Violent environments*. Ithaca & London: Cornell University Press.

Pérouse de Montclos, M.A. 2018. *L'Afrique, nouvelle frontière du djihad?* Paris: La Découverte.

Petit, V. 1997. Société d'origine et logiques migratoires. Les Dogon de Sangha (Mali). *Population* 52 (3): 515–43.

Petit, V. 1998. *Les migrations Dogon*. Paris: L'Harmattan.

Peyre de Fabrègues B., 1984. Quel avenir pour l'élevage au Sahel ? *Rev. Elev. Méd. Vét. Pays Trop.* 37 (4): 500–508.

Ping, J. 2014. The Crisis in Mali. *Harvard International Review*. Winter 2014.

Ploeg, J.D. van der. 2013. *Peasants and the art of farming: A Chayanovian manifesto*. Bourton on Dunsmore: Fernwood Publishing.

Point Carbon and Perspectives. 2008. CDM Due Diligence. Idete Reforestation Project in Tanzania. Prepared for the Ministry of Finance, Norway. Oslo.

Poulton, R.E. and I. Ag Youssouf 1998. *A peace of Timbuktu. Democratic governance, development and African peacemaking*. New York & Genève: UNIDIR.

Pulido, L. and J. De Lara. 2018. Reimaging 'justice' in environmental justice: Radical ecologies, decolonial thought, and the Black Radical Tradition. *Environment and Planning E. Nature and Space* 1 (1–2): 76–98.

Ragin, C. and H. Becker (eds). 1992. *What is a Case? Exploring the Foundations of Social Inquiry*. Cambridge: Cambridge University Press.

Ranger, T. O. 1983. The invention of tradition in colonial Africa in Hobsbawm, E. and Ranger, T. O. (eds) *The invention of tradition*. Cambridge: Cambridge University Press.

Refseth, T. 2010. Norwegian carbon plantations in Tanzania: Towards sustainable development? Master thesis, Norwegian University of Life Sciences.

Reij, C., G. Tappan and B. Belemvire. 2005. Changing land management practices and vegetation in the Central Plateau of Burkina Faso (1968–2002). *Journal of Arid Environments* 63 (3): 642–59.

Ribot, J.C. 1995. Review of Policies in the Traditional Energy Sector. Forestry Sector Policy. Washington DC: World Bank.

Ribot, J. C. 2010. Vulnerability does not just come from the sky: Framing grounded pro-poor cross-scale climate policy. *Social Dimensions of Climate Change: Equity and Vulnerability in a Warming World. New Frontiers of Social Policy, Washington, DC: The World Bank.*

Richards, P. 2004. Controversy over recent West African wars: An agrarian question? Occasional Paper, Centre for African Studies, University of Copenhagen (January 2004).

Richards, P. (ed). 2005. *No peace, no war. An anthropology of contemporary armed conflicts.* Oxford: James Currey.

Robbins, P. 2012. *Political Ecology: A Critical Introduction.* Chichester: Wiley-Blackwell.

Roberts, R., 1996. *Two Worlds of Cotton. Colonialism and the Regional Economy in the French Sudan, 1800–1946.* Stanford: Stanford University Press.

Rochegude, A., 1977, Tendances récentes du droit de la terre en République du Mali, *Revue International de Droit Comparé* 29 (1): 721–46.

Roe, E. 1991. Development narratives, or making the best of blueprint development. *World Development* 19 (4): 287–300.

Roe, E. 1999. *Except-Africa. Remaking development, rethinking power.* New Brunswick: Transaction Publishers.

Rondeau, Ch. 1980. La société sénoufo du Mali Sud (1870–1950). De la « tradition » à la dépendance. Doctoral dissertation. Université de Paris VII, Département d'Histoire.

Said, E. 1978. *Orientalism: Western Conceptions of the Orient.* London: Penguin.

Sandford, S. 1983. *Management of pastoral development in the third world.* New York: John Wiley.

Sangaré, B. 2018. Le centre du Mali: Vers une question peule? In: De Bruijn, M. (ed). *Biographies de la Radicalisation: Des messages cachés du changement social.* Bamenda, Cameroon: Langaa RPCIG. 203–224.

Sanogo, Y. 1990. Zooforé: Friend or enemy of the forest? The viewpoint of the son of a Malian peasant. Issues Paper No 15, International Institute for Environment and Development.

Schlosberg, D. 2003. The Justice of Environmental Justice: Reconciling Equity, Recognition, and Participation in a Political Movement. In A. Light and A. de-Shalit (eds), *Moral and Political Reasoning in Environmental Practice*, Cambridge MA: MIT Press.

Schlosberg, D. and L.B. Collins. 2014. From environmental to climate justice: climate change and the discourse of environmental justice. *WIREs Climate Change* 5: 359–74.

Scientific American. 2019. Massive Forest Restoration Could Greatly Slow Global Warming. 4th July. https://www.scientificamerican.com/article/massive-forest-restoration-could-greatly-slow-global-warming

Scoones, I. 2021. Pastoralists and peasants: Perspectives on agrarian change. *Journal of Peasant Studies* 48 (1): 1–47.

Scoones, I., Leach, M. and Newell, P., 2015. *The politics of green transformations.* London: Routledge.

Scott, J.C. 1976. *The moral economy of the peasants: Rebellion and subsistence in Southeast Asia.* New Haven: Yale University Press.

Scott, J.C. 1985. *The weapons of the weak: Everyday forms of peasant resistance.* New Haven: Yale University Press.

Scott, J.C. 1990. *Domination and the arts of resistance: Hidden transcripts.* New Haven: Yale University Press.

Scott, J.C. 1998. *Seeing like a state: How certain schemes to improve the human condition have failed.* New Haven: Yale University Press.

Sedgwick, M. 2015. Jihadism, Narrow and Wide: The Dangers of Loose Use of an Important Term. *Perspectives on Terrorism* 9 (2): 34–41.

Selby, J. and C. Hoffmann. 2014. Rethinking climate change, conflict and security. *Geopolitics* 19: 747–56.

Selby J, G. Daoust and C. Hoffmann. 2022. *Divided Environments: An International Political Ecology of Climate Change, Water and Security.* Cambridge: Cambridge University Press.

Shanmugaratnam, N., T. Vedeld, A. Mossige and M. Bovin. 1991. Resource management and pastoral institution building in the West African Sahel. World Bank Discussion Paper.

Shaw, S. 2013. Fallout in the Sahel: The geographic spread of conflict from Libya to Mali. *Canadian Foreign Policy Journal* 19 (2): 199–210.

Sinclair A.R.E. and J.M. Fryxell. 1985. The Sahel of Africa: Ecology of a disaster. *Canadian Journal of Zoology* 63 (5): 987–94.

Skelton, A. et al. 2020. 10 myths about net zero targets and carbon offsetting, busted. *Climate Home News,* https://www.climatechangenews.com/2020/12/11/10-myths-net-zero-targets-carbon-offsetting-busted/

Smolski, A.R. and M. Lorenzen, 2021. Introduction: Violence, Capital Accumulation, and Resistance in Contemporary Latin America. *Latin American Perspectives* 48 (1).

Soares, Benjamin. 2013. Islam in Mali since the 2012 coup. Hot Spots, *Fieldsights,* June 10. [Accessed 16.4.2024]. https://culanth.org/fieldsights/islam-in-mali-since-the-2012-coup

Solomon, H. 2013. Mali: West Africa's Afghanistan. *The RUSI Journal* 158 (1): 12–19.

Spear, T. 2003. Neo-traditionalism and the limits of invention in British colonial Africa. *Journal of African History* 44: 3–27.

Stebbing, E.P. 1937. The threat of the Sahara. *Journal of the Royal African Society* 37: 3–35.

Stott, P. and S. Sullivan (eds) 2000. *Political ecology: Science, myth and power.* London: Arnold.

Sullivan, S and R.F. Rohde 2002. On non-equilibrium in arid and semi-arid grazing systems. *Journal of Biogeography* 29: 1–26.

Sûret-Canale, J. 1962. *French colonialism in Tropical Africa, 1900–1945.* London: C. Hurst & Co.

Svarstad, H. and T.A. Benjaminsen. 2017. Nothing succeeds like success narratives: a case of conservation and development in the time of REDD. *Journal of Eastern African Studies* 11 (3): 482–505.

Svarstad, H. and T.A. Benjaminsen. 2020. Reading radical environmental justice through a political ecology lens. *Geoforum* 108: 1–11.

Svarstad, H. 2009. Narrativitetens Sosiologi. *Sosiologi i dag* 39 (4): 29–56.

Swift, J. 1996. Desertification: narratives, winners and losers. In: *The Lie of the Land: Challenging Received Wisdom on the African Environment*. Leach, M., Mearns R. (eds). London: James Currey, 73–90.

Taussig, M. 1987. *Shamanism, Colonialism, and the Wild Man. A Study in Terror and Healing*. Chicago: University of Chicago Press.

Thébaud, B and S. Batterbury. 2001. Sahel Pastoralists: Opportunism, Struggle, Conflict and Negotiation. A Case Study from Eastern Niger. *Global Environmental Change* 11: 69–78.

The Guardian. 2019. July 4. https://www.theguardian.com/environment/2019/jul/04/planting-billions-trees-best-tackle-climate-crisis-scientists-canopy-emission

Theisen, O.M., N.P. Gleditsch and H. Buhaug. 2013. Is climate change a driver of armed conflict? *Climatic Change* 117 (3): 613–25.

Thiam, A. 2017. Centre du Mali: Enjeux et dangers d'une crise négligée. Centre pour le dialogue humanitaire, Institut du Macina. Bamako, March 2017.

Thompson, E.P. 1971. The moral economy of the English crowd in the eighteenth century. *Past & Present* 50: 76–136.

Thompson, E.P. 2015. *Customs in Common: Studies in Traditional Popular Culture*. New York: New Press.

Thurston, A. 2017. *Boko Haram: The history of an African jihadist movement*. Princeton: Princeton University Press.

Toulmin, C. 2020. *Land, investment and migration. Thirty-five years of village life in Mali*. Oxford: Oxford University Press.

Touré, B. 2022. *Aménagement agricole et pastoralisme en zone office du Niger au Mali, une étude anthropologique des dynamiques foncières et des conflits liés aux ressources naturelles*. Bamako: La Sahélienne.

Tucker, C.J., Dregne, H.E. and Newcomb, W.W. 1991. Expansion and Contraction of the Sahara Desert from 1980 to 1990. *Science* 253: 299–301.

Turner, D.M, D.K. Davis, E.T. Yeh, P. Hiernaux, E.R. Loizeaux, E.M. Fornof, A.M. Rice and A.K. Suiter. 2023. Great Green Walls: Hype, Myth, and Science. *Annual Review of Environment and Resources* 48: 263–87.

Turner, M. 1992. Living on the edge: Fulbe herding practices and the relationship between economy and ecology in the inland Niger Delta of Mali. PhD dissertation, University of California, Berkeley.

Turner, M. 1993. Overstocking the range: A critical analysis of the environmental science of Sahelian pastoralism. *Economic Geography* 69 (4): 402–21.

Turner, M. 2004. Political ecology and the moral dimensions of 'resource conflicts': The case of farmer–herder conflicts in the Sahel. *Political Geography* 23: 863–89.

Turner, M. 2006. The micropolitics of common property management on the Maasina floodplains of central Mali. *Canadian Journal of African Studies* 40 (1): 41–75.

Turner, M.D., T. Carney, L. Lawler, J. Reynolds, L. Kelly, M.S. Teague and L. Brottem. 2021. Environmental rehabilitation and the vulnerability of the poor: The case of the Great Green Wall. *Land Use Policy* 111: 105750.

UNCCD 2020. The Great Green Wall Implementation Status and Way Ahead to 2030. Bonn: United Nations Convention to Combat Desertification (accessed 15 October 2021).

UNEP. 2011. Towards a Green economy: Pathways to sustainable development and poverty eradication. Nairobi: UNEP.

URT (United Republic of Tanzania). 1999. The village land Act #5. Dar es Salaam: URT.

Veldman, J.W., J.C. Aleman, S.T. Alvarado, T.M. Anderson, S. Archibald, W.J. Bond, T.W. Boutton et al. 2019. Comment on 'The Global Tree Restoration Potential.' *Science* 366 (6463).

Voice of America. 2019. 14 July. https://learningenglish.voanews.com/a/how-to-fight -climate-change-plant-a-trillion-trees/4993147.htm

Walker, P. 2006. Political ecology: where is the policy? *Progress in Human Geography* 30 (3): 382–95.

Walle, van de, N. 2001. *African economies and the politics of permanent crisis, 1979– 1999.* Cambridge: Cambridge University Press.

Walther, O. and D. Christopoulos. 2015. Islamic terrorism and the Malian rebellion. *Terrorism and Political Violence* 27 (3): 497–519.

Watts, M. 1983. *Silent Violence: Food, Famine and Peasantry in Northern Nigeria.* Berkeley: University of California Press.

Welzer, H. 2012. *Climate wars. Why people will be killed in the 21st century.* Cambridge: Polity.

Wilkens, J. and A.R.C. Datchoua-Tirvauday. 2022. *International Affairs* 98 (1): 125–43.

Wing, S.D. 2016. French Intervention in Mali: Strategic Alliances, Long-Term Regional Presence? *Small Wars & Insurgencies* 27 (1): 59–80.

Wolf, E. 1969. *Peasant wars in the 20th century.* Chapel Hill: University of Oklahoma Press.

World Bank. 2012. Inclusive Green growth: The pathway to sustainable development. Washington, D.C. World Bank.

World Bank. 2023. Document d'évaluation du projet sur un prêt d'un montant de 138,100,000 millions d'euros à la république du Mali pour un projet de restauration des terres dégradées, Mali. N° de rapport: PAD4822. Washington D.C. World Bank.

World Commission on Environment and Development (WCED). 1987. *Our common future.* Oxford: Oxford University Press.

Yeo, N.R. 2018. Diagnostic environnemental de la mise en œuvre du programme Grande Muraille Verte au Sénégal: Étude comparée des bases opérationnelles de Mbar Toubab, Koyli Alpha et Widou Thiengoly, Master's dissertation, AgroParisTech.

Young, I. 1990. *Justice and the Politics of Difference.* Princeton, NJ: Princeton University Press.

Zenn, J. 2017. Demystifying al-Qaida in Nigeria: Cases from Boko Haram's Founding, Launch of Jihad and Suicide Bombings. *Perspectives on Terrorism* 11 (6): 174–90.

Zounmenou, D. 2013. The National Movement for the Liberation of Azawad Factor in the Mali Crisis. *African Security Review* 22 (3): 167–74.

Index